The Peach-Potato Aphid (*Myzus persicae*)

Considered as economically significant pests with worldwide distribution, aphids feed on hundreds of cultivated and ornamental plants and cause considerable economic loss on a global scale.

The peach-potato aphid, *Myzus persicae*, is one of the most severe model aphid pests and successfully colonizes hundreds of plant species from 40 different families. It is a complex aphid pest known for its high reproductive rate, biotype formation, suppression of plant defenses, and wide range of host plants. This book covers ecological aspects – including life cycle, damage, and host range – which make aphids an economically important pest. A wide range of topics are covered, including ecological aspects of *M. persicae*; challenges with management approaches; damage; suppression of plant defenses; evolution; the mechanism of resistance against major classes of insecticides; the potential of biological control and natural plant-derived compounds (plant defense elicitors) as an alternative to the use of synthetic chemical insecticides to control *M. persicae*; and integrated pest management for *M. persicae*. As the peach-potato aphid is a model aphid pest, studying it helps to develop control strategies against the entire group of aphids.

Written for professionals, as well as students, teachers, and researchers in the fields of entomology, ecology, and bioscience, this book is an enriching new addition to its field.

The Peach-Potato Aphid
(*Myzus persicae*)
Ecology and Management

Jamin Ali

CRC Press
Taylor & Francis Group
Boca Raton London New York

CRC Press is an imprint of the
Taylor & Francis Group, an **informa** business

First edition published 2024
by CRC Press
2385 NW Executive Center Drive, Suite 320, Boca Raton FL 33431

and by CRC Press
4 Park Square, Milton Park, Abingdon, Oxon, OX14 4RN

CRC Press is an imprint of Taylor & Francis Group, LLC

ISBN: 978-1-032-50929-7 (hbk)
ISBN: 978-1-032-51073-6 (pbk)
ISBN: 978-1-003-40097-4 (ebk)

DOI: 10.1201/9781003400974

Typeset in Times
by SPi Technologies India Pvt Ltd (Straive)

Dedication

This book is dedicated to my late father, Al-Haj Sharafat Ali (1926–2004), and my late brother, Al-Haj Ayyoob Ali (1972–2022). Their unwavering support paved the path to my achievements and success.

Contents

Preface

More than 5000 species of aphids have been identified under 703 genera, out of which 250 species are considered as economically significant pests with worldwide distribution. Aphids feed on hundreds of crop plants and cause considerable economic loss globally. The peach-potato aphid, *Myzus persicae*, is a common and widespread pest of numerous plant species, including fruit trees, vegetables, and ornamental plants. This aphid species is known for its ability to rapidly reproduce and transmit plant viruses, causing significant economic damage to crops worldwide. Due to its high degree of adaptability and its resistance to insecticides, the management of *M. persicae* is a constant challenge for farmers and scientists. Thus, understanding the biology, ecology, and behavior of this pest is essential to develop effective and sustainable management strategies.

This monograph on the peach-potato aphid is the culmination of years of research by experts in the field. The main objective of this book is to provide a comprehensive overview of the biological, ecological, and evolutionary aspects of *M. persicae* and to explore various strategies for its management. The chapters in this book cover various aspects of the peach-potato aphid, including its life cycle, feeding behavior, aphid–plant interactions, host plant selection, transmission of plant viruses, and more. The author has brought together their experiences and expertise to provide up-to-date and critical information on this important pest.

In all, the book contains seven chapters covering various aspects of the peach-potato aphid *M. persicae*. The first chapter provides a brief introduction to the aphid as a crop pest, and special emphasis is given on aphid–plant interactions, including host plant selection, damage, and examples of some economically important aphid species. Chapters 2 and 3 cover several aspects of *M. persicae* such as life cycle, distribution, economic importance, and pest characteristics that affect its control strategies. Chapter 4 describes current management practices as well as their limitations and possible solutions. In Chapter 5, the author has compiled a few success stories in managing this pest. The final two chapters (6 and 7) are devoted to the role of resistant cultivars and induced defenses in the successful management of *M. persicae*, respectively.

This book is intended for researchers, students, and professionals interested in understanding the ecological and evolutionary aspects of aphid–plant interactions and in developing strategies for the management of *M. persicae*. The author firmly believes, and is confident, that this book will serve as a valuable resource for those involved in pest management and crop protection.

Acknowledgments

The publication of this book was only possible with the help of several people to whom I want to express my sincerest gratitude. First and foremost, I would like to thank my supervisor, Prof. Toby Bruce, for all the help, support, and encouragement he has given me over the past few years. Thanks to all my past and present colleagues and friends are also due for their support and encouragement throughout the writing of this monograph – especially to Dr. Mohammad Mukarram, Dr. Mohd Umar Farukh, Dr. Mohd Adil, Dr. Shahzar Ali Khan, Dr. Gulzar-ul-Hasan, Dr. Noorul Shajar, Dr. Mogeda Abdel Hafez, Dr. Mohd Arshad, Dr. Faizi Weqar, Syed Shane Alam, Babu Saddam Malik, and Tabish Ali Hashmi, for potentially listening to me talk about this monograph on too many occasions. Special thanks to Mr. Shuaib Anwar and his family for the delicious food during the manuscript writing.

I also want to thank Prof. Khalid Rahman Hakeem (King Abdulaziz University, Saudi Arabia), Prof. Ahmet Bayram (Dicle University, Türkiye), and Dr. Sajad Mir (SKUAST, Kashmir, India) for reviewing the book proposal and also for their suggestions that much improved chapters. I thank Alice Oven, Senior Editor at CRC Press / Taylor & Francis, who guided me during the publication process through many steps taken towards the achievement of my goal. I am also grateful to Amelia Bashford and Shikha Garg of Taylor & Francis.

Finally, I would like to extend my warmest appreciation and love to my parents (the late Haji Sharafat, the late Shapiyan, Shakeela), brothers (Haji Shabbir, the late Haji Ayyub, Hatim Ali, and Mahboob Ali), sisters (Shabbiran, Shahnaz, and Sheeba), wife (Nasra) and son (Wajdaan Zamin Sharafat) for their help, support, love, constant encouragement, and assistance throughout the years of hard work.

Author

Dr. Jamin Ali is an applied insect chemical ecologist whose expertise lies at the interface of ecology, biology, and chemistry and who focuses on using insect- and plant-derived chemicals to develop novel tools for the sustainable management of phytophagous insect pests in global agricultural environments. Dr. Ali has recently completed his PhD from Keele University, UK. Before this, he completed his master's degree in Zoology (Entomology) from Aligarh Muslim University, India. He has successfully qualified for several national-level competitive exams such as CSIR-UCGC Junior Research Fellowship (JRF) and GATE and has been awarded with several national and international fellowships. Dr. Ali has an excellent record of publications in top-notch journals. He has also presented his research work at national and international conferences and serves as a reviewer for several international journals.

Acronyms

ANOVA	Analysis of variance
BABA	ß-aminobutyric acid
BF	Blank formulation
BwYV	Beet western yellows viruses
BYV	Beet yellows viruses
CAEP	Centre of Applied Entomology and Parasitology
CaMV	Cauliflower mosaic virus
CJ	*cis*-Jasmone
CMV	Cucumber mosaic virus
Cont.	Control
CSPs	Chemosensory proteins
D	Dark
DEE	Diethyl ether
DMNT	(*E*)-4,8-dimethyl-1,3,7-nonatriene
EAG	Electroantennogram
EGBN	English giant *Brassica napus*
EPF	Entomopathogen fungi
EPNs	Entomopathogen nematodes
EPVs	Entomopathogen virus
ESTs	Expressed sequence tags
FID	Flame ionization detector
GC-MS	Gas chromatography – Mass spectrometry
GLM	Generalized linear model
h	Time (hours)
H_2O_2	Hydrogen peroxide
HIPVs	Herbivore-induced plant volatiles
IPM	Integrated pest management
JA	Jasmonic acid
L	Light
LCV	Leaflet curl virus
LD	Light dark
LMV	Lettuce mosaic virus
MeJA	Methyl jasmonate
MeSA	Methyl salicylate
MHO	6-methyl-5-hepten-2-one
ND	Not detected
OBPs	Odor-binding proteins
OPDA	Oxophytodienoic acid
ORNs	Olfactory receptor neurones
PCs	Principal components
PDJ	Prohydrojasmon
PLV	Potato leafroll virus

PTFE	Polytetrafluoroethylene
PVY	Potato virus Y
RH	Relative humidity
RI	Retention index
ROS	Reactive oxygen species
SA	Salicylic acid
SAR	Systemic acquired resistance
SBN	Samurai *Brassica napus*
SE	Standard error of the difference
TMTT	(*E,E*)-4,8,12-Trimethyl-1,3,7,11-tridecatetraene
TMV	Tobacco mosaic virus
ToYTV	Tomato yellow top virus
TR	Turnip rutabaga
TSV	Tomato spotted wilt virus
TuYV	Turnip yellow virus
VOCs	Volatile organic compounds
WMV	Watermelon mosaic viruses

1 Aphid as a Crop Pest

1.1 INTRODUCTION

Aphids (Hemiptera; Aphididae) are small, soft-bodied insects that feed on plant sap. They are among the most destructive crop pests, causing significant damage to crops and reducing crop yields (van Emden and Harrington, 2017). Aphids can infest a wide range of crops, including vegetables, fruits, cereals, and ornamental plants (Holman, 2009). Aphids reproduce rapidly, with many species capable of producing multiple generations in a single growing season (van Emden and Harrington, 2007). This means that infestations can quickly escalate, causing significant damage to crops. Aphids damage plants in several ways that affect plants' growth and development and make them nutritionally weaker (van Emden and Harrington, 2017). Among the most common symptoms of aphid attack are: stunted growth, leaf curling, distorted leaves and the deposit of a sticky residue on leaf surfaces called honeydew (Koirala, 2020).

Overall, aphids pose a significant threat to global crop production. This chapter will delve into the various aspects of aphid infestation and management. It will explore the types of crops that are susceptible to aphid attack, and the different species of aphids that can cause damage. Additionally, the chapter will examine the impact of aphids on crop yield and quality, as well as the economic implications of their presence. Furthermore, the current chapter will also discuss a few important aphid species which are serious threats to agricultural and horticultural production.

1.2 ECONOMIC IMPORTANCE

Aphids feed on plant sap using their piercing-sucking mouthparts (Abbot, Tooker and Lawson, 2018). They are known to cause significant economic damage to many crops, including cereals, vegetables, and fruits, by reducing plant growth, spreading plant viruses, and reducing the yield (Blackman and Eastop, 2000). Various studies have estimated the yield loss caused by aphids in different crops. For example: In wheat, aphids can reduce yield by up to 50% if left untreated (Kauppi *et al.*, 2021). In potato, aphids can cause yield losses ranging from 10 to 80% (Xu and Gray, 2020), depending on the severity of the infestation (Kroschel *et al.*, 2020). In soybeans, aphids can reduce yield by up to 40%, with losses of up to 100% in severe cases (Ragsdale *et al.*, 2007; Koch *et al.*, 2018). Overall, the yield loss caused by aphids in agriculture can vary depending on many factors, such as the type of crop, the severity of the infestation, and the timing of the control measures. Therefore, it is important to monitor the crops regularly and apply appropriate control methods to minimize the damage caused by aphids (Figure 1.1).

DOI: 10.1201/9781003400974-1

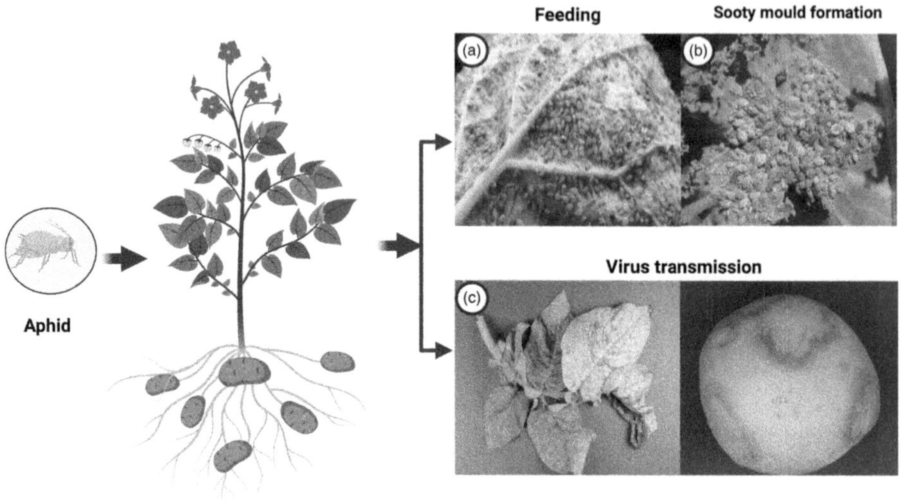

FIGURE 1.1 Aphid damages; (a) Brassica plant showing direct feeding damage (b) Sooty mould formation on brassica plant leaf due to honeydew deposition, and (c) Potato plant and tuber showing indirect damage due potato virus transmission by aphid.

1.3 ECOLOGY

Aphids are tiny insects that can cause significant damage to plants; they play essential roles in the ecosystems through their interactions with other organisms. Ecological aspects, their relationships with plants, polymorphism and mode of damage, are critical for understanding aphid ecology and the roles they play in ecosystems. One of the most crucial ecological aspects for aphids is their relationship with plants. Aphids feed on the sap of plants, and some species have specific host plants. Therefore, the availability and quality of host plants play a vital role in the growth and survival of aphid populations. Moreover, aphids also have complex symbiotic relationships with bacterial endosymbionts. These interactions can affect the behavior, survival, and reproduction of aphids. This section briefly explains a few ecological aspects of aphids and how these aspects influence their management strategies.

1.3.1 Host Plant Selection

Aphids feed on a wide range of host plants, varying from crops and ornamental plants to woody plants and weeds (Blackman and Eastop, 2000, 2006; Holman, 2009). Some of the most common hosts are *Vegetable crops*: cabbage, lettuce, tomato, pepper, cucumber, and potato (Capinera, 2020); *Fruit trees*: apple, pear, peach, and plum (Mahendiran, Lal and Sharma, 2022); *Ornamental plants*: rose, chrysanthemum, petunia, and daisy (Drees, 1993; Capinera, 2001), *Woody plants*: oak, maple, birch, and willow (van Emden and Harrington, 2007; Sorensen, 2009), and *Weeds*: milkweed, clover, and nettle (Tamaki, 1975). The host plant range for aphids varies depending on the species of aphids and the geographical location. Some aphid species

TABLE 1.1
Showing Some Key Aphids of Economic Importance along with Their Primary Host Plants

Common Name	Scientific Name	Host Plant
Peach-potato aphid	*Myzus persicae*	Peach, plum, nectarine, almond, apricot, grape, potato, lettuce, cucumber, and many other plants
Black Bean Aphid	*Aphis fabae*	Broad beans, chickpeas, peas, and other legumes.
Potato Aphid	*Macrosiphum euphorbiae*	Potato, tomato, pepper, eggplant, and other solanaceous plants
Cotton Aphid	*Aphis gossypii*	Cotton, melons, cucumbers, pumpkins, and other cucurbits
Rose Aphid	*Macrosiphum rosae*	Roses and other ornamental plants
Cabbage Aphid	*Brevicoryne brassicae*	Cabbage, broccoli, cauliflower, kale, and other cruciferous vegetables
Melon Aphid	*Aphis gossypii*	Melons, cucumber, pumpkins, and other cucurbits
Oak Aphid	*Myzocallis spp.*	Oak and other tree species
Citrus Aphid	*Toxoptera citricida*	Citrus trees, including oranges, lemons, and grapefruits
Woolly Apple Aphid	*Eriosoma lanigerum*	Apple and other fruit trees

are monophagous, meaning that they feed on a single plant species, while others are polyphagous and can feed on a wide range of plants. A list of few important aphid species with their preferred host plant is given in Table 1.1.

Host plant selection by aphids is one of the most complex procedures as the host plant has a direct impact on their reproductive health and development. The process of host selection in aphids involves a series of successive behavioral events (Powell, Tosh and Hardie, 2006; Döring, 2014) which includes: (1) Pre-alighting behavior, (2) Initial plant contact (first aphid-plant surface contact and assessment of plants before inserting the stylets), (3) Probing the epidermis layer, (4) Stylet pathway activity, (5) Sieve element puncture, and (6) Sap ingestion. As stated above, aphids have piercing-sucking mouthparts, which they use to penetrate plant tissues and extract sap. These mouthparts consist of a long, flexible proboscis, which is made up of four slender, needle-like stylets. The two inner stylets are used to puncture the plant tissue and inject saliva, while the outer stylets form a guiding sheath and act as stabilizers (Powell, Tosh and Hardie, 2006). Aphids feed by inserting their proboscis into the plant phloem, the layer of tissue that transports sap from the leaves to the rest of the plant. They then suck up the sugary sap, which is rich in nutrients, using a special pump-like organ called the cibarium. The aphids also inject saliva into the plant as they feed, which contains enzymes that break down the plant tissues and make it easier for them to extract nutrients (Mutti *et al.*, 2008). Overall, aphids are highly specialized feeders, and their mouthparts and feeding style allow them to efficiently extract sap from plants and reproduce quickly.

1.3.2 Polymorphism in Aphids

One fascinating aspect of aphids is their polymorphism, which refers to the variation in their morphology and behavior within a single species (Dixon, 1977). Polymorphism in aphids refers to the ability of individual aphids to produce offspring with different morphological forms or castes, even though they have similar genotypes (Lees, 1966; Braendle *et al.*, 2006). This phenomenon allows aphids to adapt to different environmental conditions, such as changes in temperature, humidity, and plant host (Chen *et al.*, 2019). The most common forms of polymorphism in aphids are winged and wingless forms, as well as different color morphs. Winged forms are better adapted to disperse to new host plants, while wingless forms are better adapted to reproduce on a single, stable host. Color morphs allow aphids to blend in with the color of the host plant and avoid predation (Balog, 2013). The ability of aphids to switch between different morphological forms is known as phenotypic plasticity, and it is a critical adaptation that allows aphids to thrive in different environments (Görür, 2000).

1.3.3 Aphid-Endosymbiont Microbes

Aphids are known to harbor endosymbiont microbes within their bodies, which help the aphids to survive and reproduce (Guo *et al.*, 2017). The endosymbionts are bacteria that live within the specialised cells of the aphid's body, and provide a range of benefits to the insect. The most well-known endosymbiont of aphids is *Buchnera sp.*, which is found in all aphids (Heyworth and Ferrari, 2015). *Buchnera sp.* is essential for the aphid's survival, as it provides the insect with essential amino acids that it cannot synthesize itself (Hansen and Moran, 2011). This is because the aphid's diet of plant sap is low in these amino acids. In exchange, the aphid provides a protected habitat and a consistent supply of nutrients which permit *Buchnera* to thrive. In addition to *Buchnera*, many aphids also harbor other endosymbionts, such as *Hamiltonella*, *Regiella*, and *Serratia* (Ayoubi *et al.*, 2020). These endosymbionts play a role in the aphid's immune system, helping to protect the insect from pathogens and parasites (Ayoubi *et al.*, 2020). They also produce vitamins and other nutrients that the aphid needs to survive.

The relationship between aphids and their endosymbionts is an example of mutualism, a situation in which both the host and the symbiont benefit from the relationship (Feldhaar, 2011). The endosymbionts provide essential nutrients and protection for the aphids, while the aphids provide a stable environment and a source of nutrients for the endosymbionts. Overall, the relationship between aphids and endosymbiont microbes is essential for the survival and reproduction of these insects. Without their endosymbionts, aphids would not be able to obtain the essential amino acids and other nutrients they need from their diet of plant sap.

1.4 DAMAGE

Aphids cause significant damage to plants by feeding directly on the plant tissues and depositing honeydew on the leaves (van Emden and Harrington, 2017). Direct feeding by aphids can cause leaf curling, stunted growth, and even the death of the plant.

They also weaken the plant by sucking the sap, which can lead to an infestation of other harmful insects (van Emden and Harrington, 2007, 2017). Honeydew is a sugary substance that aphids secrete while feeding on plants. This sticky substance can attract ants, fungal growth, and mold, which can further damage the plant (Hoffmann, 2016). Honeydew deposition can also lead to reduced photosynthesis in the leaves, causing yellowing and premature decay (Schwartzberg and Tumlinson, 2014; Hoffmann, 2016). Overall, aphid damage caused by direct feeding and honeydew deposition can have serious consequences for plant health and growth.

Aphids are an excellent vector for virus transmission from infected plants to healthy plants as they feed on them (Harris and Maramorosch, 2014). Some examples of aphid-transmitted key plant viruses are given here, *Tomato spotted wilt virus* (TSV): This virus infects a wide range of plants and is transmitted by several species of aphids (Hayahi *et al.*, 2018). The TSV causes a variety of symptoms, including yellowing, stunting, and the necrosis of leaves and stems (Parrella *et al.*, 2003). *Cucumber mosaic virus* (CMV): This virus can infect over 1200 plant species and is transmitted by many different aphid species (Palukaitis *et al.*, 1992; Mochizuki and Ohki, 2012). The CMV causes chlorotic mottling, distortion, and the necrosis of leaves and stunted growth (Palukaitis *et al.*, 1992). *Potato virus Y* (PVY): This virus infects potatoes and other related plants and is transmitted by several aphid species (Blanchard *et al.*, 2008). It can cause mosaic patterns on leaves, stunted growth, and reduced yield (Abbas and Amrao, 2017). Damage caused by aphid-transmitted viruses to plants can be severe and result in reduced yield, stunted growth, and even the death of the plant (Harris and Maramorosch, 2014). Moreover, the economic impact caused by these viruses is enormous. To prevent the spread of plant viruses, it is important to control aphid populations either through biological or chemical means.

1.5 APHID–PLANT INTERACTION

The interaction between aphid and plant is complex and involves various mechanisms. Aphids cause damage to plants in several ways (van Emden and Harrington, 2017). For instance, aphids feed on plant sap, which can cause stunted growth, the yellowing of leaves, and a decrease in plant productivity. Aphids excrete a sticky substance called honeydew, which can attract other insects and fungi, causing additional damage to the plant (Hoffmann, 2016; Singh and Singh, 2021). Furthermore, aphids transmit plant viruses, which can be devastating to crops (van Emden and Harrington, 2017). To defend against aphids, plants have evolved several mechanisms (Züst and Agrawal, 2016). Some plants can produce toxic compounds that repel or kill aphids. Other plants produce chemicals that attract aphid predators, such as ladybugs and lacewings. Additionally, plants can produce physical barriers, such as trichomes (small hair-like structures), that can physically prevent aphids from feeding on plant sap (Nalam, Louis and Shah, 2019). In some cases, aphids and plants can also engage in mutualistic relationships. Aphids can secrete a sugary substance called honeydew, which can attract ants. Therefore, some plants have evolved to tolerate aphids and allow them to feed on their sap in exchange for the protection that the ants provide (Auclair, 1963). In doing so, the aphids provide a food source for the ants, while the ants protect the plant from other herbivores. Overall, the interaction between aphids and plants is complex and can have far-reaching consequences on plant health and productivity.

1.5.1 Plant Defense Responses against Aphid

In an attempt to cope with aphid feeding, plants have evolved complex defense mechanisms, ranging from physical barriers to chemical defenses, induced defenses, and predator attraction as follows: (1) *Physical defenses*: Some plants have physical structures or traits that may act as physical barriers to protect themselves against aphid attack (Guerrieri and Digilio, 2008). For example, trichomes, hairs, and thorns on leaves, stems, and fruits may deter aphids by making it difficult for them to feed or reproduce (Hanley *et al.*, 2007). Additionally, some plants have thicker leaf cuticles or bark layers that can resist aphid penetration (Milenković *et al.*, 2014). (2) *Chemical defenses*: Plants produce a wide range of secondary metabolites, often referred to as phytochemicals, that play crucial roles in plant defense (Hare, 1992). These chemicals can be toxic or distasteful to aphids or act as signals to attract predators that prey on aphids. For example, terpenoids, phenolics, alkaloids, and glucosinolates are all potent plant secondary metabolites that can deter aphids from feeding or laying eggs (Saroj *et al.*, 2020). Some of these compounds, such as nicotine and pyrethrins, have insecticidal properties that can kill aphids. (3) *Induced defenses*: Plants have the ability to induce defense responses in response to aphid feeding through signal transduction pathways. This process is known as systemic acquired resistance (SAR) or induced systemic resistance (ISR).

The SAR is mediated by salicylic acid (SA) signaling, which triggers the expression of defense genes, resulting in the production of defense-related metabolites (Durrant and Dong, 2004). ISR, on the other hand, is mediated by jasmonic acid (JA) signaling (Choudhary, Prakash and Johri, 2007), which activates a different set of defense genes, resulting in the production of volatile organic compounds that can attract predators to control aphids (Serteyn *et al.*, 2020). (4) *Predator attraction*: Some plant defense mechanisms are based on indirect defenses, where plant signals attract predator insects that prey on aphids (Gatehouse, 2002). For example, plants produce volatile organic compounds that attract parasitoid wasps or ladybugs, which feed on aphids. These predators not only help to control aphid populations but also provide a feedback mechanism that enhances plant defense responses (Goggin, 2007). Conclusively, plant defense responses against aphid feeding involve a combination of physical, chemical, and induced defenses, as well as indirect defenses through predator attraction. These mechanisms help plants to minimize the damage caused by aphids, thereby ensuring their survival and reproductive success.

1.6 CURRENT MANAGEMENT PRACTICES AND CHALLENGES

Aphids are one of the most common pests in crop production, and their infestations can cause significant damage to plants. To combat this pest, current management practices have been employed such as biological control, chemical control, and cultural control. Biological control methods include the use of natural predators such as ladybugs, lacewings, and parasitoid wasps to control aphid populations.

Chemical control methods involve the use of insecticides to reduce the number of aphids. However, this method can be harmful to non-target species and the environment. Cultural control practices, such as crop rotation and the pruning of infested plants, can be used to prevent aphids from infesting crops. Overall, a combination of these practices may be employed to effectively manage aphids in crop production.

Aphids are a common plant pest that can pose various challenges for farmers, gardeners, and horticulturists. One of the primary challenges associated with aphid management is the suppression of plant defense. Aphids have evolved mechanisms that allow them to suppress their host plants' defense responses, making it easier for them to feed and reproduce (Züst and Agrawal, 2016). One way in which aphids suppress plant defense is through the secretion of saliva into their host plants' body while feeding on their sap. This saliva contains various compounds that inhibit the plant's defense responses, such as the production of toxic chemicals, reactive oxygen species (ROS), and defensive proteins (Will and van Bel, 2008). Furthermore, aphids can manipulate the signaling pathways used by plants to detect and respond to herbivore attacks (Jaouannet *et al.*, 2014).

Another important challenge associated with aphid management is the development of resistance against a major class of insecticides (Bass *et al.*, 2014). Due to the excessive use of insecticides, aphids have developed resistance to them. Therefore, it is crucial to apply a variety of control measures, including biological control, chemical treatments, and cultural practices, to manage populations effectively.

1.7 APHID SPECIES OF MOST AGRICULTURE IMPORTANCE

This section explains a few economically important aphids that cause significant damage and responsible for a considerable loss of global yield. A few important aphids have been listed with their images in Figure 1.2).

FIGURE 1.2 Showing different species of aphids which are considered as key pests for several economically important crops.

1.7.1 Pea Aphid (*Acyrthosiphon pisum*)

Pea aphid (Acyrthosiphon pisum), also known as green pea aphid, is an important insect pest in the agricultural sector, especially for pea cultivation. The species is found throughout the world and has a wide range of hosts, including legumes, especially peas, and other crops such as alfalfa and clover. Pea aphids feed on the plant sap by piercing the phloem tissue of the host plants with their stylet mouthparts. Their feeding causes stunted growth, chlorosis, and wilting of the plant, which can severely affect the plant's yield and quality. Pea aphids are also known to transmit viruses to the host plants, damaging them further. The viruses transmitted by pea aphids include pea seed-borne mosaic virus, pea yellow mosaic virus, and pea mosaic virus, which can lead to reduced crop yield and quality.

There are several cultural, biological, and chemical control methods that can be used to manage pea aphids. The cultural control methods include crop rotation, plant and row spacing, and delayed planting, all of which can reduce pea aphid populations. The use of insecticides is often the most common and cost-effective method for controlling pea aphid populations. However, over-reliance on insecticides can lead to the development of resistant populations, and the use of insecticides can also harm beneficial organisms and non-target species. Therefore, it is essential to use integrated pest management (IPM) strategies that combine various control methods, including biological and cultural methods, to control pea aphids effectively.

1.7.2 Black Bean Aphid (*Aphis fabae*)

The black bean aphid (*Aphis fabae*) is an economically important pest insect due to its ability to cause significant crop damage to a variety of plants. This pest has several common names, including the blackfly, bean aphid, and beet leaf aphid. The black bean aphid is native to Europe, but it has also been introduced into North and South America, Africa, and Asia. It prefers mild, humid weather conditions, and can survive winter in some areas. It is a small insect with an oval-shaped body, measuring around 2 mm in length. The adult aphids are usually black/grey in color, but their coloration can vary depending on the host plant and seasonal factors.

The black bean aphid is a significant pest of many vegetables, fruits, and ornamental plants. Its feeding on the sap of a plant reduces the photosynthetic rate and causes a reduction in the plant's yield, leading to significant economic losses. This pest has a wide range of host plants, including beans, beetroot, potato, tomato, cabbage, lettuce, and also fruit trees such as apple, pear, and peach. It can also infest many ornamental plants, such as roses and chrysanthemums. In conclusion, the black bean aphid is an economically important pest insect that can cause significant damage to a variety of crops and ornamental plants. It is distributed across the world, and its high reproductive capacity and resistance to insecticides make it a very challenging pest. Therefore, agricultural practices, such as crop rotation, biological control, and the use of insect-resistant varieties, have all been employed to minimize its impact.

1.7.3 Cotton Aphid (*Aphis gossypii*)

The cotton aphid (*Aphis gossypii*) is a serious pest in cotton cultivation worldwide and is commonly called melon or cotton aphid due to its preference for feeding on these crops. This species is economically important because it causes damage to crops by feeding on plant sap and by transmitting plant viruses. The cotton aphid causes substantial yield losses in cotton crops, with severe infestations causing stunted growth, leaf curling, and the yellowing or wilting of leaves, leading to a reduction in photosynthesis and reduced yield. Moreover, aphids excrete honeydew, which encourages the growth of sooty mold, a fungal disease that obstructs photosynthesis and reduces plant vigor. In addition to causing reductions in yield, cotton aphids can transmit multiple viral diseases, which can result in further economic losses. Cotton aphids have a worldwide distribution and are found in most cotton-growing regions. However, environmental factors, such as temperature, humidity, and host availability, can affect their population dynamics and distribution.

1.7.4 Mustard Aphid (*Lipaphis pseudobrassicae*)

The mustard aphid, scientifically known as *Lipaphis pseudobrassicae*, is a type of aphid that belongs to the family Aphididae. The mustard aphid is found in many parts of the world, including Europe, North America, and Asia. In addition, this pest has also been reported in other parts of Asia, including India, Pakistan, Iran, and China. It is a common pest of cruciferous crops, particularly mustard, broccoli, cauliflower, cabbage, and radish. It causes damage to host plants by removing sap from the leaves, stems, and buds. Like other aphids, this pest also sucks sap from the plant, which causes unsightly yellow patches on leaves, stunts growth, and leads to a reduced crop yield. As mentioned earlier, the mustard aphid is considered economically important, as infestations of this pest can lead to a reduction in crop yield, which can have a negative impact on the income of farmers. In addition, the use of pesticides to control these pests can be expensive and have negative impacts on the environment.

1.7.5 Potato Aphid (*Macrosiphum euphorbiae*)

The potato aphid (*Macrosiphum euphorbiae*) is a common pest of potato and other plants in the Solanaceae family such as tomato, eggplant, and bell peppers. It is commonly known as the potato aphid as it is highly destructive to the potato crop. This insect is responsible for significant economic losses in agriculture due to both direct and indirect damage of host plants. *M. euphorbiae* pierces the plant tissue with its mouthparts and sucks the plant sap, leading to stunted plant growth, the yellowing and curling of leaves, and reduced yields. This insect also excretes large amounts of honeydew, which promotes the growth of sooty mold, resulting in the further decline of plant health. Potato aphid causes significant economic damage to potato crops, which is a valuable cash crop worldwide. In addition to potato, this insect causes economic losses on other crops such as tomato, cucumber, pepper, lettuce, and bean crops. Crop losses due to potato aphid infestation can be as high as 80% and could result in significant financial losses to farmers.

The potato aphid has a global distribution, and they are widespread throughout Europe, North America, Asia, Africa, and Oceania. In the United States, they are found in every significant tomato production area and are more common in the southern states, where the weather is warmer for a more extended period. In Europe, potato aphids are common in potato and tomato production regions, especially in the Mediterranean countries. Various methods can be used to control potato aphids, including chemical and biological control methods. The use of resistant potato varieties and cultural controls such as crop rotation and planting resistant plants can also be effective in controlling aphid infestation. In conclusion, the potato aphid is a serious economic pest of potato and other Solanaceae crops worldwide. It causes significant damage to crops, resulting in reduced yields and monetary losses to farmers.

1.7.6 Peach-Potato Aphid (*Myzus persicae*)

The peach-potato aphid *M. persicae* is a common pest that belongs to the family Aphididae. This pest is known for its diverse range of hosts, which includes a variety of crops such as potato, peach, tomato, pepper, and other vegetables, herbs, and fruits. The small size, fast reproduction rate, and ability to transmit viral plant diseases make the peach-potato aphid an economically important pest in agriculture. Damage caused by the peach-potato aphid includes stunted growth, wilting, the yellowing and distortion of leaves, decreased photosynthesis, and reduced yields. In addition to direct damage, this pest is known for transmitting plant viruses such as potato virus Y (PVY), cucumber mosaic virus (CMV), and tobacco streak virus (TSV), causing further yield losses. The economic importance of *M. persicae* can be attributed to the significant crop losses it causes. This pest is considered among the most destructive aphids globally and has been reported to cause losses, ranging from 10% to 80%, in potato and other crops. This monograph is primarily focused on *M. persicae* and the following chapters will describe the ecological aspects, current management strategies, and proposed solutions for this pest in detail.

1.7.7 Bird Cherry-Oat Aphid (*Rhopalosiphum padi*)

The bird cherry-oat aphid (*Rhopalosiphum padi*) belongs to the family Aphididae. It is an important pest of cereal crops and grasses, including wheat, barley, oats, rye and also Prunus species, including bird cherry, blackthorn, and wild cherry. The species is named after its host plant, the bird cherry, which is commonly affected by this aphid. The bird cherry-oat aphid is found in many parts of the world, including Europe, Asia, and North America. In Europe, the aphids are distributed from Norway to Spain, and from the United Kingdom to Russia. In Asia, the species is also found in many countries, including China, Japan, and Korea. In North America, the aphids are distributed from Alaska to the north-eastern United States. The bird cherry-oat aphid is a highly damaging pest of cereals, especially wheat and barley. The aphids feed by piercing and sucking the sap from the plant, which can cause stunting, the yellowing of leaves, and reduced grain yield. In addition, the aphid infestation can leave a sticky excretion layer known as honeydew, which provides a favorable environment for the growth of fungi such as sooty mold. This can further reduce the yields and quality of the grain.

Moreover, the aphids are potential vectors of viral diseases that can significantly reduce crop yields. Its infestation causes significant yield losses, which can have a direct impact on the profitability of the grower. Moreover, the use of insecticides to control this aphid can also add to production costs and have an impact on the environment. In conclusion, the bird cherry-oat aphid is a significant pest of cereal crops and is widely distributed across many parts of the world. The damage caused by the aphid can significantly affect the crop yields, and effective management is necessary for sustainable crop production. The integration of various control methods can help in the effective management of aphid populations. Therefore, further research is required to develop sustainable and environmentally friendly strategies for managing bird cherry-oat aphids.

1.7.8 Grain Aphid (*Sitobion avenae*)

The *Sitobion avenae*, otherwise known as the grain aphid, is a common pest of wheat and other cereal crops throughout the world. This aphid belongs to the family Aphididae, which is one of the largest families of insects, containing thousands of species. The distribution of grain aphids is worldwide, with infestations reported in many countries. They are particularly prevalent in regions with mild winters and a temperate climate, which allows them to survive and reproduce year-round. In North America, the grain aphid is found from Canada to Mexico and they are common in the wheat-growing areas of the Great Plains. In Europe and Asia, this species is widespread and is considered a major pest of wheat crops. The grain aphids are known to infest a wide variety of host plants, although they are most commonly associated with wheat, barley, and oat crops. They have also been found on rice, sugarcane, and turf grass. This versatility in host plants allows them to survive and thrive in a variety of environments.

The damage caused by grain aphid feeding is generally associated with stunted growth, reduced yields, and poor grain quality. Severe infestations can lead to wilting, discoloration, and even the death of the plants. The economic importance of the grain aphid stems from its ability to reproduce rapidly and to infest large areas of wheat fields. Additionally, these aphids are resistant to some insecticides, making them difficult to control. Infestations of these aphids can significantly reduce wheat yields and quality, leading to billions of dollars in losses each year. In conclusion, this is an economically important pest of wheat and other cereal crops worldwide. Its ability to reproduce rapidly, infest a variety of host plants, and resist some insecticides make it difficult to control. Infestations can lead to significant reductions in yield and quality, making it a priority pest for wheat growers.

1.8 CONCLUSION

In conclusion, aphids are a significant crop pest that causes severe economic losses in agriculture. Their interaction with the host plants is complex and multifaceted, as they exploit plant defenses and manipulate plant physiology to their advantage.

Furthermore, aphid endosymbionts play a crucial role in their biology and ecology. The use of aphid pheromones as a possible control strategy has shown potential, but challenges remain in terms of their practical application. The control of aphids in agriculture is a significant challenge, requiring an integrated approach that includes biological, cultural, and chemical control methods. Successful implementation of these strategies will depend on our understanding of the ecological and biochemical interactions between the aphids, their host plants, and other organisms, as well as the development of innovative and sustainable control technologies. Overall, research on aphids as crop pests remains a critical area of investigation for agricultural sustainability, food security, and economic development.

REFERENCES

Abbas, A. and Amrao, L. (2017) 'Potato virus Y: an evolving pathogen of potato worldwide', *Pakistan Journal of Phytopathology*, 29(1), pp. 187–191.

Abbot, P., Tooker, J. and Lawson, S. P. (2018) 'Chemical ecology and sociality in aphids: opportunities and directions', *Journal of Chemical Ecology*, 44(9), pp. 770–784. doi: 10.1007/s10886-018-0955-z.

Auclair, J. L. (1963) 'Aphid feeding and nutrition', *Annual Review of Entomology*, 8(1), pp. 439–490.

Ayoubi, A. *et al.* (2020) 'Coinfection of the secondary symbionts, Hamiltonella defensa and Arsenophonus sp. contribute to the performance of the major aphid pest, Aphis gossypii (Hemiptera: Aphididae)', *Insect Science*, 27(1), pp. 86–98.

Balog, A. (2013) 'Jumping–ship–can have its costs: implications of predation and host plant species for the maintenance of pea aphid (Acyrthosiphon pisum Harris) colour polymorphism', *Bulletin of Entomological Research*, 103(5), pp. 578–583.

Bass, C. *et al.* (2014) 'The evolution of insecticide resistance in the peach potato aphid, Myzus persicae', *Insect Biochemistry and Molecular Biology*, 51(1), pp. 41–51. doi: 10.1016/j.ibmb.2014.05.003

Blackman, R. L. and Eastop, V. F. (2000) *Aphids on the world's crops: an identification and information guide*. John Wiley & Sons Ltd.

Blackman, R. L. and Eastop, V. F. (2006) *Aphids on the World's Herbaceous*. John Wiley & Sons Ltd.

Blanchard, A. *et al.* (2008) 'Potato virus Y: a century of evolution', *Virology*, 7, pp. 21–32.

Braendle, C. *et al.* (2006) 'Wing dimorphism in aphids', *Heredity*, 97(3), pp. 192–199.

Capinera, J. (2020) *Handbook of vegetable pests*. Academic Press.

Capinera, J. L. (2001) *Green Peach Aphid, Myzus persicae (Sulzer)(Insecta: Hemiptera: Aphididae)*. University of Florida Cooperative Extension Service, Institute of Food and Agricultural Sciences, EDIS.

Chen, Y. *et al.* (2019) 'Differential wing polyphenism adaptation across life stages under extreme high temperatures in corn leaf aphid', *Scientific Reports*, 9(1), p. 8744.

Choudhary, D. K., Prakash, A. and Johri, B. N. (2007) 'Induced systemic resistance (ISR) in plants: mechanism of action', *Indian Journal of Microbiology*, 47, pp. 289–297.

Dixon, A. F. G. (1977) 'Aphid ecology: life cycles, polymorphism, and population regulation', *Annual Review of Ecology and Systematics*, 8(1), pp. 329–353.

Döring, T. F. (2014) 'How aphids find their host plants, and how they don't', *Annals of Applied Biology*, 165(1), pp. 3–26. doi: 10.1111/aab.12142.

Drees, B. M. (1993) *Aphid management*. A & M University.

Durrant, W. E. and Dong, X. (2004) 'Systemic acquired resistance', *Annual Review of Phytopathology*, 42, pp. 185–209.

Feldhaar, H. (2011) 'Bacterial symbionts as mediators of ecologically important traits of insect hosts', *Ecological Entomology*, 36(5), pp. 533–543.
Gatehouse, J. A. (2002) 'Plant resistance towards insect herbivores: a dynamic interaction', *New Phytologist*, 156(2), pp. 145–169.
Goggin, F. L. (2007) 'Plant–aphid interactions: molecular and ecological perspectives', *Current Opinion in Plant Biology*, 10(4), pp. 399–408.
Görür, G. (2000) 'The role of phenotypic plasticity in host race formation and sympatric speciation in phytophagous insects, particularly in aphids', *Turkish Journal of Zoology*, 24(1), pp. 63–68.
Guerrieri, E. and Digilio, M. C. (2008) 'Aphid-plant interactions: a review', *Journal of Plant Interactions*, 3(4), pp. 223–232.
Guo, J. *et al.* (2017) 'Nine facultative endosymbionts in aphids. A review', *Journal of Asia-Pacific Entomology*, 20(3), pp. 794–801.
Hanley, M. E. *et al.* (2007) 'Plant structural traits and their role in anti-herbivore defence', *Perspectives in Plant Ecology, Evolution and Systematics*, 8(4), pp. 157–178.
Hansen, A. K. and Moran, N. A. (2011) 'Aphid genome expression reveals host–symbiont cooperation in the production of amino acids', *Proceedings of the National Academy of Sciences*, 108(7), pp. 2849–2854.
Hare, J. D. (1992) 'Herbivore-natural enemy', *Plant resistance to herbivores and pathogens: ecology, evolution, and genetics*, Editors: Fritz, R. S., Simms, E. L., University of Chicago Press, p. 278.
Harris, K. F. and Maramorosch, K. (2014) *Aphids as virus vectors*. Elsevier.
Hayahi, S. *et al.* (2018) 'Field survey of ranunculus mild mosaic virus, tomato spotted wilt virus and cucumber mosaic virus infections in Ranunculus asiaticus L. in Japan by newly developed multiplex RT-PCR', *European Journal of Plant Pathology*, 150, pp. 205–212.
Heyworth, E. R. and Ferrari, J. (2015) 'A facultative endosymbiont in aphids can provide diverse ecological benefits', *Journal of Evolutionary Biology*, 28(10), pp. 1753–1760.
Hoffmann, K. H. (2016) 'Aphid honeydew: rubbish or signaler', *Biology and ecology of aphids*, Editor: Andreas Vilcinskas, pp. 199–220.
Holman, J. (2009) *Host plant catalog of aphids: palaearctic region*. Springer.
Jaouannet, M. *et al.* (2014) 'Plant immunity in plant–aphid interactions', *Frontiers in Plant Science*, 5, p. 663.
Kauppi, K. *et al.* (2021) 'Impact of pests on cereal grain and nutrient yield in boreal growing conditions', *Agronomy*, 11(3), p. 592.
Koch, R. L. *et al.* (2018) 'Management of insecticide-resistant soybean aphids in the Upper Midwest of the United States', *Journal of Integrated Pest Management*, 9(1), p. 23.
Koirala, S. (2020) 'Mustard aphid and crop production', *International Journal of Applied Sciences and Biotechnology*, 8(3), pp. 310–317.
Kroschel, J. *et al.* (2020) 'Insect pests affecting potatoes in tropical, subtropical, and temperate regions', in *The potato crop: Its agricultural, nutritional and social contribution to humankind*, Editors: Hugo Campos, Oscar Ortiz, Springer Cham, pp. 251–306.
Lees, A. D. (1966) 'The control of polymorphism in aphids', in *Advances in insect physiology*, Editor: J.W.L. Beament, Elsevier, pp. 207–277.
Mahendiran, G., Lal, S. and Sharma, O. C. (2022) 'Pests and their management on temperate fruits: (apple, pear, peach, apricot, cherry, persimmon, walnut, olive, kiwifruit and strawberry)', in *Trends in horticultural entomology*, Editor: M. Mani, Springer Singapore, pp. 891–941.
Milenković, S. *et al.* (2014) 'Strawberry resistance to the aphid Chaetosiphon fragaefolii Cockerell (Homoptera: Aphididae)', *Pesticides and Phytomedicine/Pesticidi i fitomedicina*, 29(4).
Mochizuki, T. and Ohki, S. T. (2012) 'Cucumber mosaic virus: viral genes as virulence determinants', *Molecular Plant Pathology*, 13(3), pp. 217–225.

Mutti, N. S. *et al.* (2008) 'A protein from the salivary glands of the pea aphid, Acyrthosiphon pisum, is essential in feeding on a host plant', *Proceedings of the National Academy of Sciences of the United States of America*, 105(29), pp. 9965–9969. doi: 10.1073/pnas.0708958105.

Nalam, V., Louis, J. and Shah, J. (2019) 'Plant defense against aphids, the pest extraordinaire', *Plant Science*, 279, pp. 96–107.

Palukaitis, P. *et al.* (1992) 'Cucumber mosaic virus', *Advances in Virus Research*, 41, pp. 281–348.

Parrella, G. *et al.* (2003) 'An update of the host range of Tomato spotted wilt virus', *Journal of Plant Pathology*, pp. 227–264.

Powell, G., Tosh, C. R. and Hardie, J. (2006) 'Host plant selection by aphids: behavioral, evolutionary, and applied perspectives', *Annual Review of Entomology*, 51, pp. 309–330.

Ragsdale, D. W. *et al.* (2007) 'Economic threshold for soybean aphid (Hemiptera: Aphididae)', *Journal of Economic Entomology*, 100(4), pp. 1258–1267.

Saroj, A. *et al.* (2020) 'Phytochemicals of plant-derived essential oils: A novel green approach against pests', in *Natural remedies for pest, disease and weed control*. Elsevier, pp. 65–79.

Schwartzberg, E. G. and Tumlinson, J. H. (2014) 'Aphid honeydew alters plant defence responses', *Functional Ecology*, 28(2), pp. 386–394.

Serteyn, L. *et al.* (2020) 'Induced systemic resistance by a plant growth-promoting rhizobacterium impacts development and feeding behavior of aphids', *Insects*, 11(4), p. 234.

Singh, R. and Singh, G. (2021) 'Aphids', in *Polyphagous pests of crops*, Editor: Omkar, pp. 105–182.

Sorensen, J. T. (2009) 'Aphids', in *Encyclopedia of insects*, Editors: Vincent H. Resh, Ring T. Cardé, Elsevier, pp. 27–31.

Tamaki, G. (1975) 'Weeds in orchards as important alternate sources of green peach aphids in late spring', *Environmental Entomology*, 4(6), pp. 958–960.

van Emden, H. F. and Harrington, R. (2007) *Aphids as crop pests*. 1st edn. Edited by H. F. van Emden and R. Harrington. CABI.

van Emden, H. F. and Harrington, R. (2017) *Aphids as crop pests*. CABI.

Will, T. and van Bel, A. J. E. (2008) 'Induction as well as suppression: How aphid saliva may exert opposite effects on plant defense', *Plant Signaling & Behavior*, 3(6), pp. 427–430.

Xu, Y. and Gray, S. M. (2020) 'Aphids and their transmitted potato viruses: A continuous challenges in potato crops', *Journal of Integrative Agriculture*, 19(2), pp. 367–375. doi: 10.1016/S2095-3119(19)62842-X.

Züst, T. and Agrawal, A. A. (2016) 'Mechanisms and evolution of plant resistance to aphids', *Nature Plants*, 2(1), pp. 1–9.

2 Peach-Potato Aphid *Myzus persicae* General Introduction

2.1 INTRODUCTION

The peach-potato aphid, *Myzus persicae* Sulzer (Hemiptera; Aphididae), is a highly polyphagous and cosmopolitan pest with widespread global distribution, including significant portions of North America, Europe and Asia (van Emden *et al.*, 1969; Ali, 2022). *M. persicae* feeds on more than 500 plants from 40 different families, and spread hundreds of plant viruses (Holman, 2009; Xu and Gray, 2020). It has a very short generation time and high reproductive rate and reaches a density sufficient to cause noticeable crop injuries in a short period of time (Capinera, 2001, 2012, 2020). A high density of aphids causes a reduction in growth and water stress which, in turn, results in the yield reduction and causes further damage (Capinera, 2020). During the process of feeding, it releases saliva containing effector proteins which overwhelm the plant defense system by modifying the normal cell functioning inside the host plant (Hogenhout *et al.*, 2009).

In addition to direct damage to host plants, *M. persicae* also causes serious harm to plant health by transmitting plant viruses (persistent and non-persistent), and is considered to be an important vector for plant viruses, where both nymphs and adults are responsible for transmission (Namba and Sylvester, 1981; Eigenbrode *et al.*, 2002; Qi *et al.*, 2021). The presence of winged morphs plays a prominent role in spreading diseases due to their flying ability and dispersive nature (Mauck *et al.*, 2012). Furthermore, *M. persicae* also excretes a considerable amount of sugary liquid (honeydew), which affects the health of the host plant and also its interaction with other herbivores (Fiebig, Poehling and Borgemeister, 2004; Hoffmann, 2016).

Overall, the peach-potato aphid *M. persicae* is an important pest that causes significant damage to several agricultural crops across the globe. As such, understanding the ecological and environmental factors that influence its population dynamics can provide insights into pest management strategies that are both effective and sustainable. This chapter will delve into the various ecological aspects associated with the peach-potato aphid, including its life cycle, distribution, feeding, host range and polyphenism, with the aim of providing a comprehensive overview of its ecology to develop the sustainable ways for its management.

2.2 LIFE CYCLE

The life cycle of *M. persicae* varies according to local climatic conditions, principally depending on the severity of the winter. Van Emden *et al.* (1969) reviewed the life

DOI: 10.1201/9781003400974-2

cycle and found that development can be varied: it may be rapid with over 20 generations a season in a moderate climate, each having been completed in 10–12 days. Based on experiments in Greece, regional variations in the life cycle of *M. persicae* has been reported (Margaritopoulos *et al.*, 2002). In the case of the unavailability of the summer host plant and cold climate, aphids will overwinter on *Prunus* spp during the egg stage. In the spring, the plant emerges from the dormancy stage and starts to grow, eggs hatch and nymphs begin to feed on flowers, stems and young foliage. In a cold climate, adults return to *Prunus* spp. in the autumn, where mating happens, and eggs are laid (Figure 2.1).

When they first emerge, the aphid eggs are yellow green in colour, but they soon turn black. They are elliptical in shape and their size ranges from 0.3 mm to 0.6 mm; a high rate of mortality is recorded at this stage (Capinera, 2012). The egg hatches into a nymph, which is initially greenish but become yellowish and resemble viviparous (parthenogenetic, nymph-producing) adults. There are two assumptions about the number of developmental stages of *M. persicae*. According to Horsfall (1924), *M. persicae* has four instars in its developmental stage with the duration of each averaging 2.0, 2.1, 2.3, and 2.0 days, respectively. While according to MacGillivray and Anderson (1958), *M. persicae* has a life cycle with five instar stages with a mean developmental time of 2.4, 1.8, 2.0, 2.1 and 0.7 days, respectively. However, Horsfall (1924) has reported four instars with a mean developmental time 2.0, 2.1, 2.3 and 2.0, respectively.

Adult *M. persicae* can occupy up to eight generations on its winter host prunus, but due to the high density alate (winged) morphs are produced that migrate on to summer hosts. Alate are entirely different to the apterous (wingless) form of *M. persicae*, it has a black head and thorax, and a yellowish abdomen with dark patches on the dorsal side; the body length ranges from 1.8 to 2.1 mm (Capinera, 2001; Jung *et al.*, 2021). Alates are economically more important because of their dispersive nature and they colonize a huge number of plants, serving as an effective viral vector for plants. Later alate forms produce apterous offspring on the overwintering host plant and produces young ones (van Emden *et al.*, 1969; Capinera, 2001; Mauck, De Moraes and Mescher, 2010).

According to their host, aphids exhibit two different types of life cycle: host alternating (heteroecious) and non-host alternating (monoecious). Heteroecious aphids live on woody plants (the primary host) in winter, and migrate to unrelated secondary non-woody (herbaceous) plants in the summer. They return to the primary host when conditions become favourable again. Monoecious aphids, by contrast, usually live on the same host plant species or closely related species throughout the year (van Emden and Harrington, 2007; Hardie, 2017). Aphid generations interrupt their parthenogenesis with sexual mode of reproduction are referred as holocyclic.

In contrast, anholocyclic aphids continually reproduce parthenogenetically. In the case of holocyclic aphids, after meeting the male, the female lays its eggs on the primary winter host. Some aphid species exhibit both holocycly and anholocyly in different clonal species. *M. persicae* shows anholocyly and heteroecious behaviour, thereby circumventing the need for host alternation (Hardie, 2017). Host alternating aphids are economically more important because of the host alternation on secondary hosts, which are generally herbaceous and crop plants. Similarly, alate morphs are responsible for infestation at longer distance than typically observed among the apterous (Hardie, 2017).

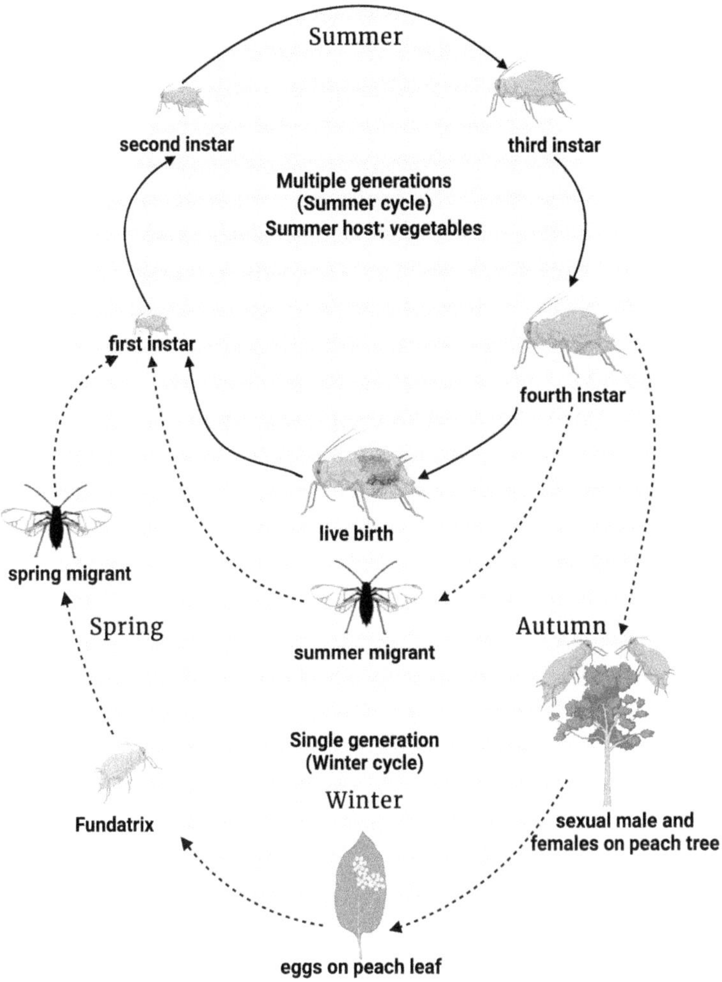

FIGURE 2.1 Illustration showing a life cycle of *M. persicae* in the presence of its host peach tree (winter) and vegetables (summer). In the winter part of the life cycle, the sexual forms of the male and female aphid *M. persicae* mate on the host plant and produce eggs to overwinter. Females (Fundatrix) hatch from the eggs and develop into Fundatrigenae (apterous aphids); these, in turn, produce winged spring migrants, which return to the secondary spring and summer host plants where they reproduce asexually. Later, after the return migration to the winter (primary) host, these daughter clones develop into winged males and females (gynoparae). The gynoparae produces the oviparae which mate with the males on the primary host, lay their eggs on the suitable places on the primary host plants.

2.3 DISTRIBUTION

The distribution of organisms is an indicator of its complexity and adaptability to a wide range of biotic and abiotic factors, the more resilient the organisms are have more diverse their distribution and consequently they generally inhabit a broad range of geographical habitats (Peterson, Allen and Holling, 1998; Guisan and Thuiller, 2005).

Aphids have been identified as cosmopolitan pests, which cause damage to plants on a global scale (van Emden and Harrington, 2017). For instance, *M. persicae* most probably has an Asian origin but it can be now be seen worldwide (Margaritopoulos *et al.*, 2009; Ali *et al.*, 2022). It is to be encountered in almost all regions from Asia to Europe, from North America to South America, and from Africa to Australia (Figure 2.2) (van Emden *et al.*, 1969; Hlaoui *et al.*, 2019). *M. persicae* has been identified as a principal pest in several regions of the world, particularly in Europe (Capinera, 2001; Cocu *et al.*, 2005; Bass *et al.*, 2015). The wide range of habitats makes this pest genetically more diverse and adaptable to different environmental conditions (Cocu *et al.*, 2005; Charaabi *et al.*, 2018).

2.4 HOST PLANT

The *M. persicae* is a highly polyphagous insect that feeds on more than hundreds of plant species belonging to 40 different families, including Brassicaceae (Blackman and Eastop, 2006; Ali, 2022), Caryophyllaceae (Edde, 2021), Cucurbitaceae (Qi *et al.*, 2021) and Fabaceae (Edde, 2021). In particular brassica, potato, pumpkin, peach, squash, lettuce, sugarbeet, sugarcane, bean, kale, melon and tobacco are the most common host plants for *M. persicae* (Capinera, 2020). It is a cosmopolitan pest that can be found on vegetables (Ali *et al.*, 2021), ornamental plants (Feng and Isman, 1995; Capinera, 2001), plants grown in the greenhouses or the field (Capinera, 2001), and that it causes severe leaf curl and shoot distortion by sap sucking (Xu and Gray, 2020).

2.5 DAMAGE

M. persicae causes damage to its host in various ways, including direct feeding damage and/or indirect damage by transmitting plant viruses that spread diseases in plants.

Direct damage; as mentioned earlier, aphids are silent feeders, they do not cause any considerable tissue damage during feeding. The peach-potato aphid penetrates the plant surface with their stylet and feeds on phloem sap (Tjallingii and Esch, 1993; Hewer *et al.*, 2011) which reduces plant vigour and causes wilting, dwarfing, the yellowing of leaves and the production of chlorotic spots (Blackman and Eastop, 2000; Bass *et al.*, 2014). Like other aphids, *M. persicae* also secretes effector proteins in their saliva at feeding site, with the help of these proteins *M. persicae* feeds for longer time without activating the plant defense responses (production of hydrogen peroxide (H_2O_2) and reactive oxygen species (ROS)).

Indirect damage; *M. persicae* is an important vector that transmits hundreds of plant viruses (Blackman and Eastop, 2000; Naga *et al.*, 2020). An increasing body of evidence reported that *M. persicae* acts as an important vector for several plant viruses. Kennedy, Day and Eastop (1962) recorded more than 100 viruses spread by *M. persicae* that include potato leafroll virus (PLV), potato virus Y (PVY), cucumber mosaic virus (CMV), beet western yellows viruses (BwYV), beet yellows viruses (BYV), lettuce mosaic virus (LMV), turnip mosaic virus (TMV), cauliflower mosaic

FIGURE 2.2 World map of areas affected by peach-potato aphid *Myzus persicae*.

virus (CaMV) and watermelon mosaic viruses (WMV). Because of its ability to transmit plant viruses *M. persicae* is considered as a severe threat to plants (Capinera, 2001; Eigenbrode *et al.*, 2002; Qi *et al.*, 2021).

Moreover, honeydew excretion also has a negative effect on plant physiology and affects the photosynthetic activities by covering the leaf surface (Zhou *et al.*, 2015). In addition, honeydew deposits serve as a source of microbial accumulation and breeding that results into plant disease and secondary infection (Fiebig *et al.*, 2004) and also the quality of crops due to its sticky features. Taken as a whole, *M. persicae* plant infestation hampers the overall growth and development of the host plant by interfering with its photosynthetic activities and chlorophyll content that indirectly affects host defense responses and make it less resistant to herbivores compared to non-infested plants (Macedo, 2003; Singh *et al.*, 2021).

2.6 FEEDING

The *M. persicae* is considered to be a silent feeder which causes significant economic losses to growers worldwide without any noticeable tissue injuries to the host plant (Cao *et al.*, 2018). The feeding process involves several steps, starting from landing on the plant surface to sap ingestion. This section will discuss the feeding steps of *M. persicae* on plants, based on the available literature (Blackman and Eastop, 2006; Will and Vilcinskas, 2013; Garzo *et al.*, 2020; Jhou *et al.*, 2021). One important point is that the following steps are not particularly associated with *M. persicae* only but also represent the feeding pattern (host plant selection) in aphids.

Step 1: *Antennal probing and testing the plant's suitability*: The first step in the feeding process of *M. persicae* is landing on the plant and antennal probing. Aphids use their antennae to detect chemical cues from their host plants, which help them to navigate to suitable feeding sites. The antennae are also used to sample the plant surface for defense-related compounds and to test the suitability of the plant for feeding.

Step 2: *Penetration into plant tissues*: Once the aphid has detected a suitable feeding site, it uses its sharp, stylus-like mouthparts to penetrate the plant tissue. The aphid's mouthparts consist of a pair of mandibles and a long, slender, flexible stylet bundle that can penetrate through the plant cells. The stylets are used to probe the plant tissue for sap, which the aphid takes up by suction.

Step 3: *Salivary secretions*: During the penetration process, aphids secrete saliva into the plant tissue. Aphid saliva contains a range of enzymes and effectors that help to facilitate feeding, suppress plant defenses, and manipulate plant physiology. The primary function of saliva is to facilitate sap uptake from the plant cells, but it can also play a role in suppressing plant defense responses, modifying plant growth and development, and transmitting plant viruses.

Step 4: *Sap ingestion*: Once the aphid's stylets have penetrated a plant cell, the aphid can begin to withdraw sap from the cell. The sap contains a mix of sugars, amino acids, and other nutrients that the aphid requires to survive and reproduce. The aphid uses suction to withdraw sap from the plant cells, which generates a negative pressure that pulls the sap into the aphid's gut.

Step 5: *Rejection of waste products*: As part of the feeding process, aphids also excrete honeydew, a sugary liquid that is produced from excess sap ingestion. Honeydew is excreted from the rear end of the aphid and can accumulate on the plant surface. Honeydew can attract other insects, such as ants, which can act as mutualistic partners for aphids.

In conclusion, the feeding steps of *M. persicae* on plants involve several complex interactions between the aphid and the plant. The process involves antennal probing, penetration into plant tissues, salivary secretions, sap ingestion, and excretion of honeydew. Understanding the feeding biology of *M. persicae* is critical for developing effective strategies for controlling this widespread pest.

2.7 POLYMORPHISM AND HOST ALTERNATION

The *M. persicae* exhibits polymorphism, meaning that different individuals within the same species have different forms or phenotypes (Dixon, 1985; Ali, 2022). This pest possesses different morphs: winged and wingless. The winged morph of *M. persicae* develops on the primary host plant (often a peach tree), where it reproduces asexually and rapidly builds up the population (Dreyer and Campbell, 1987). These winged individuals have wings and are capable of flying to secondary host plants, where they reproduce sexually and give rise to the wingless morph (Müller, Williams and Hardie, 2001; Ali, 2022). The wingless morph, by contrast, is adapted to feeding on the secondary host plant and reproduces asexually, giving rise to a large population of wingless individuals.

This strategy of alternating between sexual and asexual reproduction on different host plants is known as host alternation and it allows *M. persicae* to exploit a wide range of host plants and to rapidly adapt to changing environmental conditions (Blackman and Eastop, 2006; Holman, 2009). The ability to fly and the polymorphic nature of the species are key adaptations that facilitate this host alternation strategy (Cantó-Tejero *et al.*, 2022).

2.8 CONCLUSION

Based on these ecological aspects, the peach-potato aphid *M. persicae* take advantages of these factors to survive in a broad and diverse habitat. For instance, the high reproductive rate and short maturation helps *M. persicae* to quickly attain a population equal to the threshold level. The presence of host alternation enables this pest to survive on secondary host plants in the absence of a primary host and overcome the adverse effect of absence of a primary host which is used for food and shelter. Furthermore, *M. persicae* also exhibits polyphenism that gives rise the production of different morphs and provides several advantages to this pest, for instance; at high densities *M. persicae* produces alate (winged) morphs, which are able to fly and avoid food competition. Due to their dispersive nature, alate morphs are economically important forms of aphids they can cause infestation in distant crops.

REFERENCES

Ali, J. *et al.* (2021) 'Effects of cis-jasmone treatment of brassicas on interactions with myzus persicae aphids and their parasitoid diaeretiella rapae', *Frontiers in Plant Science*, 12. doi: 10.3389/fpls.2021.711896

Ali, J. (2022) *The chemical ecology of a model aphid pest, Myzus persicae, and its natural enemies*. Keele University.

Ali, J., Sobhy, I. S. and Bruce, T. J. A. (2022) 'Wild potato ancestors as potential sources of resistance to the aphid Myzus persicae', *Pest Management Science*, 78(9), pp. 3931–3938.

Bass, C., Puinean, A. M., *et al.* (2014) 'The evolution of insecticide resistance in the peach potato aphid, Myzus persicae', *Insect Biochemistry and Molecular Biology*, 51(1), pp. 41–51. doi: 10.1016/j.ibmb.2014.05.003

Bass, C. *et al.* (2015) 'The global status of insect resistance to neonicotinoid insecticides', *Pesticide Biochemistry and Physiology*, 121, pp. 78–87.

Blackman, R. L. and Eastop, V. F. (2000) *Aphids on the world's crops: an identification and information guide*. John Wiley & Sons Ltd.

Blackman, R. L. and Eastop, V. F. (2006) *Aphids on the world's herbaceous plants and shrubs: An identification and information guide.*. `John Wiley & Sons Ltd.

Cantó-Tejero, M. *et al.* (2022) 'Essential oils-based repellents for the management of Myzus persicae and Macrosiphum euphorbiae', *Journal of Pest Science*, 95(1), pp. 365–379.

Cao, H.-H. *et al.* (2018) 'Nutrition versus defense: why Myzus persicae (green peach aphid) prefers and performs better on young leaves of cabbage', *PloS One*, 13(4), p. e0196219.

Capinera, J. (2020) *Handbook of vegetable pests*. Academic Press.

Capinera, J. L. (2001) *Green peach aphid, myzus persicae (sulzer)(insecta: hemiptera: aphididae)*. University of Florida Cooperative Extension Service, Institute of Food and ….

Capinera, J. L. (2012) *Green peach aphid, myzus persicae (sulzer) (hemiptera: aphididae)*. Springer, pp. 1–10. doi: 10.1007/springerreference_86370

Charaabi, K. *et al.* (2018) 'Occurrence of target-site resistance to neonicotinoids in the aphid Myzus persicae in Tunisia, and its status on different host plants', *Pest Management Science*, 74(6), pp. 1297–1301.

Cocu, N. *et al.* (2005) 'Geographical location, climate and land use influences on the phenology and numbers of the aphid, Myzus persicae, in Europe', *Journal of Biogeography*, 32(4), pp. 615–632.

Dixon, A. F. G. (1985) 'Structure of aphid populations', *Annual Review of Entomology*, 30(1), pp. 155–174.

Dreyer, D. L. and Campbell, B. C. (1987) 'Chemical basis of host-plant resistance to aphids', *Plant, Cell & Environment*, 10(5), pp. 353–361.

Edde, P. A. (2021) *Field crop arthropod pests of economic importance*. Academic Press.

Eigenbrode, S. D. *et al.* (2002) 'Volatiles from potato plants infected with potato leafroll virus attract and arrest the virus vector, Myzus persicae (Homoptera: Aphididae)', *Proceedings of the Royal Society B: Biological Sciences*, 269(1490), pp. 455–460. doi: 10.1098/rspb.2001.1909

Feng, R. and Isman, M. B. (1995) 'Selection for resistance to azadirachtin in the green peach aphid, Myzus persicae', *Experientia*, 51(8), pp. 831–833.

Fiebig, M., Poehling, H. and Borgemeister, C. (2004) 'Barley yellow dwarf virus, wheat, and Sitobion avenae: a case of trilateral interactions', *Entomologia Experimentalis et Applicata*, 110(1), pp. 11–21.

Garzo, E. *et al.* (2020) 'Feeding behavior and virus-transmission ability of insect vectors exposed to systemic insecticides', *Plants*, 9(7), p. 895.

Guisan, A. and Thuiller, W. (2005) 'Predicting species distribution: offering more than simple habitat models', *Ecology Letters*, 8(9), pp. 993–1009.

Hardie, J. (2017) 'Life cycles and polyphenism', in *Aphids as crop pests*. 2nd ed. CABI, pp. 81–97.
Hewer, A., Becker, A. and van Bel, A. J. E. (2011) 'An aphid's Odyssey–the cortical quest for the vascular bundle', *Journal of Experimental Biology*, 214(22), pp. 3868–3879.
Hlaoui, A. *et al.* (2019) 'Spatial and temporal genetic diversity of the peach potato aphid Myzus persicae (Sulzer) in Tunisia', *Insects*, 10(10), p. 330.
Hoffmann, K. H. (2016) 'Aphid honeydew: rubbish or signaler', in *Biology and ecology of aphids*, Editors: Andreas Vilcinskas, pp. 199–220.
Hogenhout, S. A. *et al.* (2009) 'Emerging concepts in effector biology of plant-associated organisms', *Molecular Plant-Microbe Interactions*, 22(2), pp. 115–122.
Holman, J. (2009) *Host plant catalog of aphids: palaearctic region*. Springer.
Horsfall, J. L. (1924) *Life history studies of Myzus persicae Sulzer*. Pennsylvania State College Agricultural Experiment Station.
Jhou, Y.-S. *et al.* (2021) 'Host acceptance and plant resistance: a comparative behavioral study of Myzus persicae and Acyrthosiphon pisum', *Insects*, 12(11), p. 975.
Jung, Y. J. *et al.* (2021) 'The complete mitochondrial genome of Myzus persicae (Sulzer, 1776; Hemiptera: Aphididae) isolated in Korea', *Mitochondrial DNA Part B*, 6(1), pp. 10–12.
Kennedy, J. S., Day, M. F. and Eastop, V. F. (1962) 'A conspectus of aphids as vectors of plant viruses', in *A conspectus of aphids as vectors of plant viruses*, Commonw. Inst. Ent.
Macedo, T. B. (2003) *Physiological responses of plants to piercing-sucking arthropods*. The University of Nebraska-Lincoln.
MacGillivray, M.E. and Anderson, G.B., 1958. Development of four species of aphids (Homoptera) on potato1. *The Canadian Entomologist*, *90*(3), pp. 148–155.
Margaritopoulos, J. T. *et al.* (2002) 'Life cycle variation of Myzus persicae (Hemiptera: Aphididae) in Greece', *Bulletin of Entomological Research*, 92(4), pp. 309–319. doi: 10.1079/ber2002167
Margaritopoulos, J. T. *et al.* (2009) 'Tracking the global dispersal of a cosmopolitan insect pest, the peach potato aphid', *BMC Ecology*, 9(1), pp. 1–13.
Mauck, K. *et al.* (2012) 'Transmission mechanisms shape pathogen effects on host–vector interactions: evidence from plant viruses', *Functional Ecology*, 26(5), pp. 1162–1175.
Mauck, K. E., De Moraes, C. M. and Mescher, M. C. (2010) 'Deceptive chemical signals induced by a plant virus attract insect vectors to inferior hosts', *Proceedings of the National Academy of Sciences*, 107(8), pp. 3600–3605.
Mullens, B. A. (2019) 'Horse flies and deer flies (Tabanidae)', in *Medical and veterinary entomology*. Elsevier, pp. 327–343.
Müller, C. B., Williams, I. S. and Hardie, J. (2001) 'The role of nutrition, crowding and interspecific interactions in the development of winged aphids', *Ecological Entomology*, 26(3), pp. 330–340.
Naga, K. C. *et al.* (2020) 'Transmission efficiency of Potato leaf roll virus (PLRV) by potato aphid Aulacorthum solani and green peach aphid Myzus persicae', *Indian Journal of Entomology*, 82(1), pp. 68–71.
Namba, R. and Sylvester, E. S. (1981) 'Transmission of cauliflower mosaic virus by the green peach, turnip, cabbage, and pea aphids', *Journal of Economic Entomology*, 74(5), pp. 546–551.
Peterson, G., Allen, C. R. and Holling, C. S. (1998) 'Ecological resilience, biodiversity, and scale', *Ecosystems*, 1(1), pp. 6–18.
Qi, Y.-H. *et al.* (2021) 'Physical contact transmission of Cucumber green mottle mosaic virus by Myzus persicae', *PloS One*, 16(6), p. e0252856.
Singh, A. *et al.* (2021) 'The efficiency of plant defense: aphid pest pressure does not alter production of food rewards by okra plants in ant presence', *Frontiers in Plant Science*, 12, p. 627570.

Tjallingii, W. F. and Esch, T. H. H. (1993) 'Fine structure of aphid stylet routes in plant tissues in correlation with EPG signals', *Physiological Entomology*, 18(3), pp. 317–328.
van Emden, H. F. *et al.* (1969) 'The ecology of Myzus persicae', *Annual Review of Entomology*, 14(1), pp. 197–270.
van Emden, H. F. and Harrington, R. (2007) *Aphids as crop pests*. 1st edn. Edited by H. F. van Emden and R. Harrington. CABI.
van Emden, H. F. and Harrington, R. (2017) *Aphids as crop pests*. CABI.
Will, T. and Vilcinskas, A. (2013) 'Aphid-proof plants: biotechnology-based approaches for aphid control', in *Yellow biotechnology ii: insect biotechnology in plant protection and industry*, Editor: Andreas Vilcinskas, Springer Berlin, pp. 179–203.
Xu, Y. and Gray, S. M. (2020) 'Aphids and their transmitted potato viruses: A continuous challenges in potato crops', *Journal of Integrative Agriculture*, 19(2), pp. 367–375. doi: 10.1016/S2095-3119(19)62842-X
Zhou, S. *et al.* (2015) 'Alteration of plant primary metabolism in response to insect herbivory', *Plant Physiology*, 169(3), pp. 1488–1498.

3 Economic Importance and Traits Influence Control Strategies

3.1 INTRODUCTION

The peach-potato aphid is considered to be one of the most destructive agricultural pests and causes billions of dollars of crop losses on a global basis (Capinera, 2001). This aphid species is found throughout the world and has a wide range of host plants, including vegetables, cereals, fruit trees, and ornamental plants. *M. persicae* feeds on over 40 plant families, including Apiaceae (carrot), Asteraceae (lettuce and artichoke), Amaranthaceae (sugar beet, amaranths), Brassicaceae (broccoli, brussels sprouts, cabbage, cauliflower, kale, mustard, radish and turnip), Cucurbitaceae (cucumber, squash), Fabaceae (bean, lentils and pea), Poaceae (maize, wheat, barley and rice) and Solanaceae (potato, pepper, and tomato) (Blackman and Eastop, 2000, 2006; Holman, 2009; Bass *et al.*, 2014; Edde, 2021; Ali, 2022).

M. persicae is responsible for both qualitative and quantitative losses of agricultural production through causing chlorosis, necrosis, wilting defoliation, and even, in extremis, the destruction of the fruit. It directly affects the plant's nutritional status by feeding on plant sap (Hewer, Becker and van Bel, 2011). In addition, the peach-potato aphid acts as an important vector and transmits hundreds of plant viruses, indirectly affecting the growth and development of the host plant (Blackman and Eastop, 2000; Capinera, 2001; Eigenbrode *et al.*, 2002; Naga *et al.*, 2020; Ali *et al.*, 2021; Qi *et al.*, 2021). Moreover, *M. persicae* produces a sugary secretion (honeydew) which hampers the photosynthetic activities, plant health, fruit quality and serves as a source of microbial accumulation and breeding that results into plant disease and secondary infection (Fiebig, Poehling and Borgemeister, 2004; Zhou *et al.*, 2015). Furthermore, *M. persicae* can increase the cost of production for farmers, leading to increased consumer prices. Insect infestations can cause economic losses to farmers as they spend more money on crop protection. This can lead to the use of more expensive chemicals or more frequent applications of pesticides, which can increase the production cost of crops. As these rise, the consumer price of crops also increases. Therefore, *M. persicae* infestations have the potential to negatively impact food security, especially with regard to low-income households.

The ecological aspects of the peach-potato aphid support its status as a serious crop pest and allows it to feed on a wide range of crops (van Emden *et al.*, 1969; Margaritopoulos *et al.*, 2002). This aphid species has a complex life cycle (both sexual and asexual), leading to the telescoping of generations, a short maturation time, the presence of different morphs (both winged and wingless) and alternate host

DOI: 10.1201/9781003400974-3

plant preferences (Kindlmann and Dixon, 1989; Ali, 2022). The aphid reproduces rapidly and can cause extensive damage to crops by feeding on plant sap, thereby suppressing host defense responses (Elzinga, De Vos and Jander, 2014). It has the ability to adapt to the use of a new plant as a host plant, biotype formation, and has also evolved a resistance against several major insecticides (Bass *et al.*, 2014).

In this chapter, we discuss the economic importance of the peach-potato aphid *Myzus persicae*. We have highlighted the damage it causes to various crops and its role as a vector of plant viruses. Additionally, we have shed light on key characteristics of *M. persicae* that influence its pest status and control strategies enable it to survive in adverse conditions (non-availability of host plants) and make it one of the most persistent and disease-resistant pests for farmers and agriculturists. We have explored its reproductive capability, host plant range, and biotype formation, as well as its ability to suppress host defense and the development of resistance against insecticides. Understanding these traits is essential for the effective management of this important agricultural pest. This chapter will provide valuable insights for researchers, growers, and other stakeholders in their efforts to devise effective measures for the prevention and control of *M. persicae* infestations, ultimately leading to sustainable and profitable crop production.

3.2 ECONOMIC IMPORTANCE OF THE PEACH-POTATO APHID

As stated earlier, the peach-potato aphid is considered to be one of the most destructive agricultural pests that feeds on over 40 plant families, including Apiaceae (carrot); Asteraceae (lettuce and artichoke); Amaranthaceae (beet, spinach); Brassicaceae (broccoli, brussels sprouts, cabbage, cauliflower, kale, mustard, radish and turnip); Cucurbitaceae (cucumber, squash); Fabaceae (bean, pea, Poaceae (maize, wheat, barley and rice) and Solanaceae (potato, pepper, and tomato) (Blackman and Eastop, 2000, 2006; Holman, 2009; Bass *et al.*, 2014; Edde, 2021; Ali, 2022). *M. persicae* is responsible for both qualitative and quantitative losses of agricultural production through causing chlorosis, necrosis, wilting defoliation and the abortion of the fruit. It directly affects the plant nutritional status by feeding on plant sap (Hewer, Becker and van Bel, 2011). Additionally, the peach-potato aphid acts as an important vector and transmits hundreds of plant viruses, including potato leafroll virus (PLV), potato virus Y (PVY), beet western yellows virus (BYV), lettuce mosaic virus, cauliflower mosaic virus, turnip mosaic virus, cucumber mosaic and watermelon mosaic viruses, which indirectly affect the growth and development of the host plant (Blackman and Eastop, 2000; Capinera, 2001; Eigenbrode *et al.*, 2002; Naga *et al.*, 2020; Ali *et al.*, 2021; Qi *et al.*, 2021).

According to previous studies, *M. persicae* is responsible for a considerable loss in crop yield causing billions of dollars of global losses (Ali, 2022). Specifically, in the UK, aphid infestation has been calculated to cause up to $70 million of losses per year (Hemming *et al.*, 2022). *M. persicae* is considered to be one of the most common and devastating brassica pests in the UK, causing direct and indirect damage by spreading one of the most deleterious plant virus turnip yellow virus (TuYV), which can reduce the final yield by up to 26% (Hemming *et al.*, 2022). It has been estimated that controlling this virus could result in an increase in total yield profit equivalent to

TABLE 3.1
Showing Agricultural Loss in Several Important Crops Caused by *M. persicae*

Crop	Yield Loss (%)	Country	References
Sweet pepper	31–100	India	Sharma *et al.* (2022)
Potato	14–60	Pakistan	Khan *et al.* (2011)
Potato	up to 100	Greece	Chatzivassiliou *et al.* (2008)
Brassica	up to 92	China	Jeon *et al.* (2008)
Amaranth	up to 100	Kenya	Nampeera *et al.* (2023)
Brassica	26	UK	Hemming *et al.* (2022)
Lentils	36	Australia	Nancarrow *et al.* (2022)
Pea	45	Australia	Nancarrow *et al.* (2022)
Canola	up to 46	Australia	Nancarrow *et al.* (2022)
Tomato	20–70	Brazil	Pinheiro *et al.* (2013)
Sugar beet	11–50	Europe	Francis *et al.* (2022)

This yield loss is the result of damage caused by *M. persicae* by direct feeding and indirectly by transmitting plant viruses.

£60–90 million per year for oilseed rape growers in the UK (Stevens *et al.*, 2008). In India, the peach-potato aphid is a key pest of sweet pepper that has been observed to cause 38 to 42% yield loss under a controlled environment (Sharma, Sood and Ghongade, 2022). In Brazil, the peach-potato aphid primarily serves as a vector for PYV and tomato yellow top virus (ToYTV) viruses which cause 20–70% loss of tomato yield and reduce fruit production up to 85% (Pinheiro *et al.*, 2013). In Australia, this pest is responsible for the transmission of TuYV in pea, broad bean, lupin, chickpea, lentil crop fields and a loss of 46% canola yield has been reported (Nancarrow *et al.*, 2022). In New Zealand, *M. persicae* transmits potato virus Y (PVY) and leaflet curl virus (LCV), which are among the most damaging potato viruses and cause significant yield loss (London *et al.*, 2020). In China, *M. persicae* is also considered a primary pest for tobacco (Li *et al.*, 2016). Interestingly, in the absence of a suitable host plant *M. persicae* can easily switch to non-crop plants, including weeds such as *Convolvulus arvensis*, *Chenopodium album* and *Amaranthus retroflexes*, which act as aphid reservoirs and attack the subsequent crops (Annis, Tamaki and Berry, 1981; Nampeera, 2022). For instance, in Washington, weeds grown on the floor of peach orchards had between 3 and 16 million aphids, and it was found that around one-third of the aphids were infected with BwYV (Tamaki, 1975; Tamaki and Fox, 1982). However, the loss caused by *M. persicae* depends on plant cultivar, time of infestation, and environmental conditions (Sidauruk and Sipayung, 2018).

The *M. persicae* infestation on host plants not only causes direct damage to the plant by removing large amounts of phloem sap but also indirectly increases the susceptibility of the plant to subsequent herbivore attacks (Drizou *et al.*, 2018; Zhuang *et al.*, 2018). One of the ways that *M. persicae* makes the plant more vulnerable to subsequent herbivore attacks is by inducing changes in plant metabolism and defense mechanisms (Zhuang *et al.*, 2018). During feeding, *M. persicae* injects saliva into the plant. This contains a variety of enzymes and proteins that manipulate the plant's physiology (Harmel *et al.*, 2008). The aphid saliva interferes with the plant's signaling pathways and enzymes, triggering the production of defense compounds that are less effective against subsequent herbivory. This results in the reduced effectiveness of the plant's defense mechanisms against herbivores. In particular, *M. persicae* infestation on *Sinapis alba* reduced the concentrations of chemical volatiles produced by the plants that are known to attract natural enemies of herbivores (Pareja *et al.*, 2012). This effect resulted in fewer predatory insects around the plant, making it more vulnerable to subsequent herbivore attacks.

Another way in which *M. persicae* can increase a plant's susceptibility to subsequent herbivores is through the transmission of plant viruses (Mauck, De Moraes and Mescher, 2010). Aphids are known vectors of plant viruses, and the feeding of *M. persicae* on plants can transmit these viruses, which can cause plant stress or death. Plant viruses, such as potato virus Y, have been found to affect the levels of certain hormones that regulate plant growth and defense mechanisms (Petek *et al.*, 2014). This, in turn, can cause plants to become more susceptible to subsequent herbivore attacks.

For example, in a study by Westwood *et al.* (2014), feeding by *M. persicae* on tobacco plants infected with tobacco mosaic virus (TMV) caused the inhibition of jasmonic acid (JA)-regulated gene expression, resulting in a decrease in jasmonic acid levels. These hormonal changes in the plant resulted in increased susceptibility to insect herbivory. Conclusively, *M. persicae* can make plants more susceptible to subsequent herbivore attacks through a number of different mechanisms. These include inducing changes in plant metabolism and defense mechanisms (Zhuang *et al.*, 2018), reducing the effectiveness of plant-produced volatiles for attracting natural enemies of herbivores (Pareja *et al.*, 2012), and transmitting plant viruses that affect the plant's growth and hormonal regulation (Mauck, De Moraes and Mescher, 2010). By better understanding these mechanisms, it is possible to develop new strategies for controlling *M. persicae* and mitigating the damage they cause to crops.

3.3 TRAITS INFLUENCE THE CONTROL STRATEGIES

These ecological aspects significantly influence control techniques employed against this pest (Figure 3.1). For instance, due to biotype formation and evolution of insecticide resistance majority of the available insecticides are ineffective against this pest.

3.3.1 Reproduction and Short Maturation Time

M. persicae is considered an important pest due to its rapid adaptation to environmental changes, high reproductive rate, and short maturation time (van Emden *et al.*, 1969; Margaritopoulos *et al.*, 2002). As a result, it is essential to explore how these features affect the control strategies.

Firstly, the high reproductive rate of *M. persicae* has serious implications for control strategies. Females can produce asexual offspring, which could lead to exponential population growth. They can also produce sexual offspring that contribute to genetic diversity, allowing the species to adapt rapidly to environmental changes. Once the population of *M. persicae* reaches a high density, it can be very difficult to

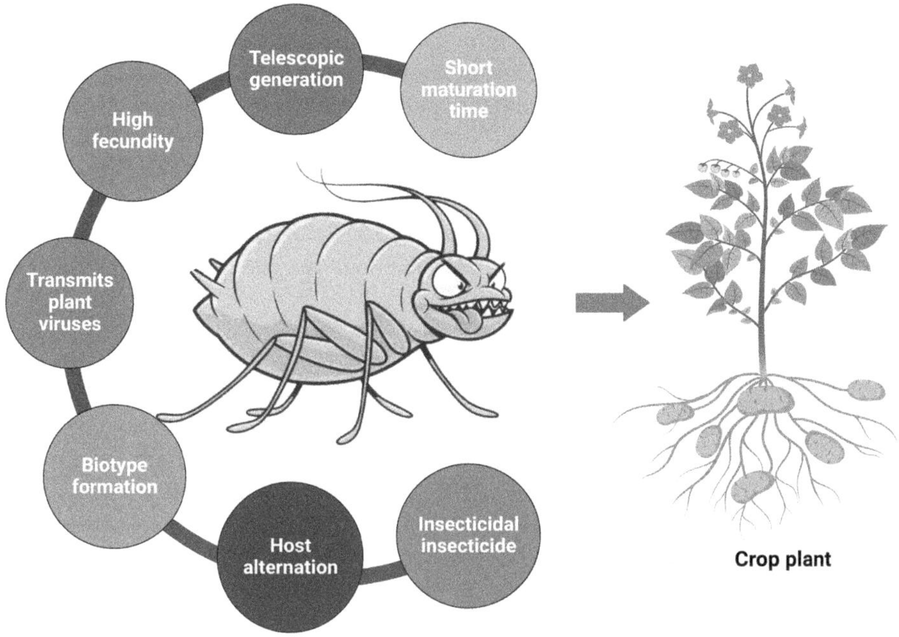

FIGURE 3.1 Factors that ensures the ecological success of *M. persicae* as a polyphagous and cosmopolitan pest of economic importance.

control. These aphids are known for their fast reproduction rate, which can lead to a rapid increase in their numbers. As the population density increases, aphids may consume more plant food sources and spread viruses or other diseases to their host plants.

Secondly, the short maturation time of *M. persicae* also poses significant challenges for control strategies. The insect reaches sexual maturity in just five days and has been documented to complete a full life cycle in less than two weeks (van Emden *et al.*, 1969). This rapid development time means that the aphids can build populations quickly, leading to significant crop damage in a short period, especially in greenhouses. However, there may be some variation in the number of days required for *M. persicae* to complete its life cycle depending on the host plant status and environmental conditions (van Emden *et al.*, 1969; Chi and Su, 2006). The successful control of *M. persicae* populations needs to be implemented as soon as possible to prevent the population from reaching an unmanageable level.

Additionally, the short life span and short maturation period makes it a challenge to develop effective biological control strategies. The biological control agents, including parasitic wasps, predators, and diseases, require a considerable amount of time to establish their population and provide effective control. However, these biological control agents still play a vital role in IPM programs. The agents can either be purchased and released on the crop, or farmers can create a conducive environment that promotes their proliferation.

Thirdly, the high reproductive rate and rapid maturation time make *M. persicae* populations resilient and persistent, providing a constant source of infection and virus transmission. The spread of viruses can have significant economic effects on agricultural production, particularly in cases where the infected crop destroys the market value of the product. In conclusion, the high reproductive rate and short maturation time of *M. persicae* pose significant challenges to the development of effective control strategies. Given the species' ability to adapt quickly to environmental changes, chemical control methods pose a challenge due to insecticide resistance. However, the use of biological control agents still holds premises. The short life span and maturation time of *M. persicae* make effective biological control difficult, and other control options, such as cultural control methods, may be crucial in reducing the pest population.

3.3.2 Telescoping of Generations

The *M. persicae* possesses another astonishing characteristic of telescoping of generation which has significant implications for the success of control strategies (Dixon, 1985). Telescopic generation refers to the phenomenon where the mother aphid has the capability to give birth to fully developed offspring, known as secondary or Fundatrix generation, that can also reproduce asexually, leading to the production of further generations (Dixon, 2012). The production of multiple generations within a single growing season increases the total number of aphids per plant greatly (Dixon, 2012; Cocuzza, 2019).

The capability of *M. persicae* to attain high population density in a short period of time amplifies the challenge of controlling this pest. In conclusion, telescopic generation in *M. persicae* poses a significant challenge to pest control strategies. The continuous generation of aphids within a single growing season implies that established control strategies require regular monitoring and revision for effectiveness.

3.3.3 Suppression of Plant Defenses

Aphids are silent feeders that causes overall less physical damage, in contrast to other herbivores, which have chewing-type mouth parts (van Emden and Harrington, 2007; Kallure *et al.*, 2022). Previous studies undertaken on *M. persicae* host–plant interactions reported that the plant has a tendency to recognise the aphid-secreted salivary proteins and modulate defense strategies which include the production of deterrent compounds (indole glucosinolates, coverts indol-3-ylmethoxyglucosinolates (I3M) to 4-methoxyindol-3-ylmethoxygluosinolates (4MI3M)), callose formation, plugging of sieve tubes and accumulation of reactive oxygen species (particularly hydrogen peroxide) (Will and van Bel, 2006; Kim *et al.*, 2008; Elzinga, Vos and Jander, 2014; Züst and Agrawal, 2016).

To overcome the plant induced-antifeeding strategies *M. persicae* releases saliva (watery and gel forms) that contains several effector proteins, including Mp1, Mp55, MpC002, Mp59, Mp60, Mp61, Mp62 and Mp63 (Hogenhout and Bos, 2011; Elzinga, Vos and Jander, 2014; Bhattacharya, 2019). In addition to saliva, aphids also excrete honeydew, which contains effector proteins that also have a negatively impact on the activation defense-related pathways in host plants (Schwartzberg and Tumlinson, 2014).

The expression of *M. persicae* salivary proteins in the host plant makes the plant more susceptible to *M. persicae* by increasing fecundity and reducing the synthesis of defensive compounds (Elzinga, De Vos and Jander, 2014). For instance, the overexpression of salivary protein Mp55 and MpC003 are responsible for high aphid fecundity, and a reduction in the production of deterrent compounds, and makes the plant more attractive to *M. persicae* in choice bioassays (Bos *et al.*, 2010; Jaouannet *et al.*, 2014). *A. thaliana* overexpressing Mp55 proteins showed high aphid fecundity, a low amount of deterrent compounds (indole glucosinolates, 4MI3M), reduced callose deposition, and a low abundance of defense signalling molecule (hydrogen molecule) (Elzinga, De Vos and Jander, 2014). The expression of *M. persicae* induced salivary protein (Mp55) presumably avoids the activation of plant defense system and promotes aphid feeding (Nalam, Louis and Shah, 2019).

Furthermore, salivary effector proteins Mp1, Mp59, Mp60, Mp61, Mp62 and Mp63 did not show any effect on aphid performance (Elzinga, De Vos and Jander, 2014). The expression of these effector proteins are host-specific and *M. persicae* modulates their activity according to the available host plants (Pitino and Hogenhout, 2013). Previous studies reported a positive co-relation between the expression of Mp55 proteins and the performance of *M. persicae* in brassica tobacco and Arabidopsis (Elzinga, De Vos and Jander, 2014; Machado-Assefh *et al.*, 2015; Silva-Sanzana *et al.*, 2022). Conclusively, effector proteins perform a broad range of the anti-defensive function in host plants such as reducing the accumulation of toxic compounds, inhibiting callose formation and lowering the production of signalling molecules (Elzinga, De Vos and Jander, 2014). It is quite clear that *M. persicae* release these compounds to suppress the defense response in host plants.

3.3.4 The Evolution of Resistance to Insecticides

One of the biggest challenges with *M. persicae* is the development of resistance to a wide range of commercially available insecticides (Nauen and Denholm, 2005; Ali *et al.*, 2021). In 1950, the first incidence of insecticide resistance in *M. persicae* was reported (Feng and Isman, 1995). The excessive use of synthetic chemical pesticides to control the peach-potato aphid over the years has evolved this species to become increasingly insecticide-resistant (Anthon, 1955; Devonshire, 1989; Soderlund and Bloomquist, 1990; Anstead *et al.*, 2005; Margaritopoulos *et al.*, 2007, 2021; Bass *et al.*, 2014). To date, *M. persicae* has developed resistance to several classes of insecticides (Anstead, Williamson and Denholm, 2005; Margaritopoulos *et al.*, 2007).

The co-evolution of insecticide use and *M. persicae* damage has remarkably advanced this species to develop mechanisms that avoid or detoxify the effect of synthetic chemicals (Bass *et al.*, 2014; Troczka *et al.*, 2021; Ali, 2022). Bass *et al.* (2014) has described the evolution of insecticide resistance into several mechanisms (Figure 3.2) that includes the induction and enhanced expression of mutation and enzymes responsible for resistance. Each induced change (overexpression of enzymes and/or mutation) is responsible for resistance to a specific class of chemicals (Bass *et al.*, 2011, 2014; Pym *et al.*, 2022).

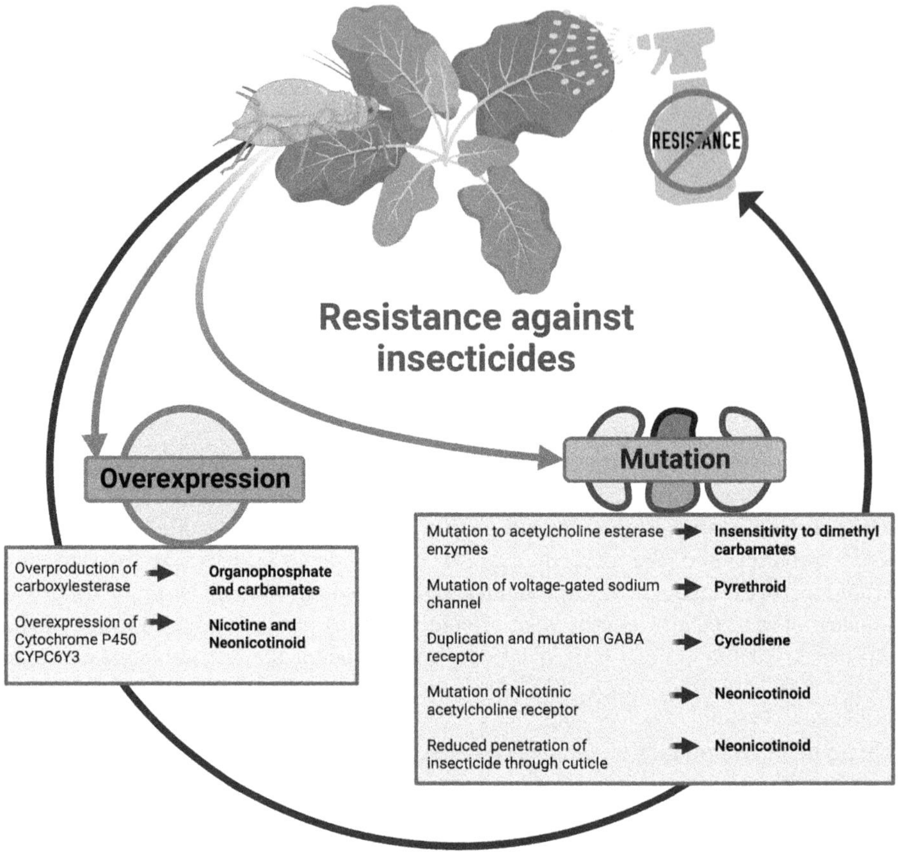

FIGURE 3.2 Illustration showing the causes responsible for the development of resistance to particular insecticides (Bass *et al.*, 2014). According to Bass *et al.* (2014), there are two main causes that play a significant role in the evolution of resistance in insects against synthetic insecticides: overexpression of genes responsible for the synthesis of detoxifying enzymes and mutation in target sites. The ability to develop and spread resistance against insecticide is also facilitated by high fecundity, short generation times, and the ability to disperse over long distances.

3.3.5 Polymorphism

As mentioned above, polymorphism refers to the existence of multiple forms or morphs within a species. *M. persicae*, commonly known as the peach-potato aphid, exhibits significant polymorphism in terms of both color and shape (Blackman and Eastop, 2000). The most common form of *M. persicae* is the green form, which is found primarily on the leaves and stems of host plants. However, there are also pink and red forms which are induced by environmental cues, such as temperature and light conditions (Margaritopoulos *et al.*, 2002). These colour morphs are thought to be adaptations to different environments and provide camouflage against predators.

In addition to colour polymorphism, *M. persicae* also exhibits morphological polymorphism. This includes the generation of both winged and wingless forms, as well as forms with varying body lengths, antennal lengths, and tarsal segment numbers (Park and Hardie, 2002; Hardie, 2017). These morphs are also thought to be adaptations to changes in environmental conditions, such as food availability or habitat quality. The polymorphism in *M. persicae* has been extensively studied due to the pest status of this species. Understanding the genetic basis of polymorphism can provide insight into the evolution and ecology of this species, as well as inform pest management strategies (Castañeda *et al.*, 2010).

The economic importance of aphid polymorphism lies in its impact on pest management strategies (Williams and Dixon, 2007). The different morphs of aphids have varying susceptibilities to different control measures, such as insecticides or biological control agents. For example, the wingless forms may be more susceptible to biological control than the winged forms, which can disperse easily and escape control measures. Furthermore, the presence of different morphs can complicate the task of monitoring and predicting pest outbreaks. In conclusion, the polymorphism of aphids has important implications for pest management and crop production. Further research is needed to fully understand the mechanisms underlying polymorphism in aphids and develop more effective pest management strategies.

3.3.6 Biotype Formation

Biotype (morphologically indistinguishable subspecies that are different in behavior or physiology) formation in *M. persicae* permits it to extend its host range and quickly occupy a new ecological niche (Tagu *et al.*, 2008). The formation of biotypes means that developing insect-resistant crop varieties is a more complicated and challenging process for this species as it can easily adapt to new resistant varieties too (Nauen and Denholm, 2005; Singh, Jasrotia and Crespo-Herreraa, 2022). In addition, it has been observed that plant resistance is frequently overcome after a few years because of the appearance of new aphid biotypes (Alvarez *et al.*, 2006). Biotype formation is a process where a population of a particular species diverges into distinct subpopulations that have different genetic and physiological traits, allowing them to adapt to different environmental conditions or to overcome resistance mechanisms of their hosts. Biotype formation is increasingly studied in insect pests, such as the peach-potato aphid *M. persicae*, which is a highly adaptable and globally distributed agricultural pest. This section will discuss the mechanisms and advantages of biotype formation in *M. persicae*, as well as the implications for pest management strategies (Simon and Peccoud, 2018; Shih, Sugio and Simon, 2023).

M. persicae has been shown to exhibit high genetic variation within and among populations, which can lead to the development of biotypes with different physiological and behavioral traits. Biotype formation in *M. persicae* can occur through several mechanisms, including: (1) *Host plant specialization: M. persicae* has a wide host range, being able to feed on over 400 plant species. However, some populations have been shown to specialize on certain host plants, which can select for specific traits that enable them to exploit those plants more efficiently (Shih, Sugio and Simon, 2023). (2) *Insecticide resistance: M. persicae* has become a major pest in agricultural systems due to its rapid adaptation to insecticides. This can lead to the emergence of insecticide-resistant biotypes that are able to survive and reproduce in the presence of insecticides that would normally kill susceptible individuals. Different biotypes can exhibit different resistance mechanisms, such as increased detoxification enzymes, target site modifications, or reduced penetration of insecticides (Daborn and Le Goff, 2004; Bass *et al.*, 2015).

The emergence of biotypes in *M. persicae* can confer several advantages that allow these insects to exploit different ecological niches and overcome the challenges of their environment. Among the advantages are: (1) *Increased fitness*: Biotypes that specialize on certain host plants or develop resistance to insecticides or viruses may have higher fitness than generalist populations or non-resistant biotypes (Bruce, 2010). This is because they are better adapted to the specific challenges of their environment, such as the presence of toxins or competition from other insects. (2) *Rapid adaptation*: The ability of *M. persicae* to produce new biotypes in response to changing environmental conditions or host plants can enable these insects to rapidly adapt to novel situations (Simon and Peccoud, 2018; Shih, Sugio and Simon, 2023). This can be particularly important in agricultural systems, where new insecticides or resistant crops can be introduced at a rapid pace, forcing pests to continually adapt to new selection pressures. (3) *Increased virulence*: In some cases, biotypes of *M. persicae* that develop resistance to insecticides or viruses may become more virulent than susceptible populations, allowing them to cause more damage to crops or spread more viruses to other plants

The emergence of biotypes in *M. persicae* poses significant challenges for pest management strategies, which often rely on the use of pesticides or resistant crops. Insecticide-resistant biotypes can greatly reduce the effectiveness of insecticide applications. Additionally, biotypes that specialize on certain crops or varieties can cause significant damage to those plants, requiring targeted management strategies that are often more costly and less effective than broad-spectrum approaches. To overcome these challenges, researchers and practitioners are developing new strategies that take into account the genetic and ecological diversity of *M. persicae* populations (Wakil, Brust and Perring, 2017; Uriel, 2020).

Conclusively, the formation of biotypes in *M. persicae* is a complex and dynamic process that can confer significant advantages in terms of adaptation, fitness, and virulence. However, it also poses significant challenges for pest management strategies, requiring more targeted and sustainable approaches that take into account. Furthermore, the formation of biotypes in *M. persicae* plays a critical role in its survival, reproduction, and adaptation to various host plants. The ability to evolve and adapt genetically to host plant resistance mechanisms helps *M. persicae* to overcome these defenses and exploit new host plants.

3.4 CONCLUSION

In conclusion, *M. persicae* is an important economic pest that has significant impacts on global agricultural systems. These insect pests cause significant yield losses by transmitting viruses and causing direct feeding damage. Furthermore, the ecological traits of *M. persicae*, including their rapid reproductive rates, their ability to form different biotypes, the presence of different morphs, the suppression of plant defense and the evolution of resistance to insecticides, all have significant implications for control strategies. The continuous evolution of *M. persicae*, driven by the selective pressure of control measures, poses a significant challenge to the management of this pest. Farmers and producers have to make significant investments in controlling pests such as aphids, including the use of chemical pesticides, which can be costly and harmful to the environment and public health. Therefore, efforts must be made to promote sustainable pest management techniques that reduce the use of conventional pesticides in favor of biological and cultural control methods. These efforts can help to reduce the economic impact of *M. persicae* on agricultural systems while promoting sustainable food production practices. Thus, an integrated pest management strategy, which includes biological, chemical, and cultural control methods, is crucial for the effective control of this pest. Furthermore, regular monitoring and continuous research are necessary to track its evolving ecological traits and develop new control strategies that are sustainable and effective.

REFERENCES

Ali, J. (2022) *The chemical ecology of a model aphid pest, Myzus persicae, and its natural enemies*. Keele University.

Ali, J. *et al.* (2021) 'Effects of cis-jasmone treatment of brassicas on interactions with myzus persicae aphids and their parasitoid diaeretiella rapae', *Frontiers in Plant Science*, 12. doi: 10.3389/fpls.2021.711896

Alvarez, A. E. *et al.* (2006) 'Location of resistance factors in the leaves of potato and wild tuber-bearing Solanum species to the aphid Myzus persicae', *Entomologia Experimentalis et Applicata*, 121(2), pp. 145–157.

Annis, B., Tamaki, G. and Berry, R. E. (1981) 'Seasonal occurrence of wild secondary hosts of the green peach aphid, Myzus persicae (Sulzer), in agricultural systems in the Yakima Valley', *Environmental Entomology*, 10(3), pp. 307–312.

Anstead, J. A., Williamson, M. S. and Denholm, I. (2005) 'Evidence for multiple origins of identical insecticide resistance mutations in the aphid Myzus persicae', *Insect Biochemistry and Molecular Biology*, 35(3), pp. 249–256.

Anthon, E. W. (1955) 'Evidence for green peach aphid resistance to organo-phosphorous insecticides', *Journal of Economic Entomology*, 48(1), pp. 56–57.

Bass, C., Puinean, A. M., *et al.* (2014) 'The evolution of insecticide resistance in the peach-potato aphid, Myzus persicae', *Insect Biochemistry and Molecular Biology*, 51(1), pp. 41–51. doi: 10.1016/j.ibmb.2014.05.003

Bass, C. *et al.* (2011) 'Mutation of a nicotinic acetylcholine receptor β subunit is associated with resistance to neonicotinoid insecticides in the aphid Myzus persicae', *BMC Neuroscience*, 12(1), pp. 1–11.

Bass, C. *et al.* (2015) 'The global status of insect resistance to neonicotinoid insecticides', *Pesticide Biochemistry and Physiology*, 121, pp. 78–87.

Bhattacharya, S. (2019) 'Brassica-aphid interaction: challenges and prospects of genetic engineering for integrated aphid management', *Physiological and Molecular Plant Pathology*, 108, p. 101442.

Blackman, R. L. and Eastop, V. F. (2000) *Aphids on the world's crops: an identification and information guide*. John Wiley & Sons Ltd.

Blackman, R. L. and Eastop, V. F. (2006) *Aphids on the world's herbaceous*. John Wiley & Sons Ltd.

Bos, J. I. B. *et al.* (2010) 'A functional genomics approach identifies candidate effectors from the aphid species Myzus persicae (green peach aphid)', *PLoS Genetics*, 6(11), p. e1001216.
Bruce, T. J. A. (2010) 'Tackling the threat to food security caused by crop pests in the new millennium', *Food Security*, 2(2), pp. 133–141. doi: 10.1007/s12571-010-0061-8
Capinera, J. L. (2001) *Green peach aphid, myzus persicae (sulzer)(insecta: hemiptera: aphididae).* University of Florida Cooperative Extension Service, Institute of Food and Agricultural Sciences, EDIS.
Castañeda, L. E. *et al.* (2010) 'Effects of wing polyphenism, aphid genotype and host plant chemistry on energy metabolism of the grain aphid, Sitobion avenae', *Journal of Insect Physiology*, 56(12), pp. 1920–1924.
Chatzivassiliou, E. K. *et al.* (2008) 'Infection of potato crops and seeds with Potato virus Y and Potato leafroll virus in Greece', *Journal of Plant Pathology*, pp. 253–261.
Chi, H. and Su, H.-Y. (2006) 'Age-stage, two-sex life tables of Aphidius gifuensis (Ashmead) (Hymenoptera: Braconidae) and its host Myzus persicae (Sulzer)(Homoptera: Aphididae) with mathematical proof of the relationship between female fecundity and the net reproductive rate', *Environmental Entomology*, 35(1), pp. 10–21.
Cocuzza, G. E. M. (2019) *Insect pests of fruits: aphids.* CABI.
Daborn, P. J. and Le Goff, G. (2004) 'The genetics and genomics of insecticide resistance', *Trends in Genetics*, 20(3), pp. 163–170.
Devonshire, A. L. (1989) 'Insecticide resistance in Myzus persicae: from field to gene and back again', *Pesticide Science*, 26(4), pp. 375–382.
Dixon, A. F. G. (1985) 'Structure of aphid populations', *Annual Review of Entomology*, 30(1), pp. 155–174.
Dixon, A. F. G. (2012) *Aphid ecology an optimization approach.* Springer Science & Business Media.
Drizou, F. *et al.* (2018) 'Infestation by Myzus persicae Increases Susceptibility of Brassica napus cv."Canard" to Rhizoctonia solani AG 2-1', *Frontiers in Plant Science*, 9, p. 1903.
Edde, P. A. (2021) *Field crop arthropod pests of economic importance.* Academic Press.
Eigenbrode, S. D. *et al.* (2002) 'Volatiles from potato plants infected with potato leafroll virus attract and arrest the virus vector, Myzus persicae (Homoptera: Aphididae)', *Proceedings of the Royal Society B: Biological Sciences*, 269(1490), pp. 455–460. doi: 10.1098/rspb.2001.1909
Elzinga, D. A., De Vos, M. and Jander, G. (2014) 'Suppression of plant defenses by a Myzus persicae (green peach aphid) salivary effector protein', *Molecular Plant-Microbe Interactions*, 27(7), pp. 747–756.
Feng, R. and Isman, M. B. (1995) 'Selection for resistance to azadirachtin in the green peach aphid, Myzus persicae', *Experientia*, 51(8), pp. 831–833.
Fiebig, M., Poehling, H. and Borgemeister, C. (2004) 'Barley yellow dwarf virus, wheat, and Sitobion avenae: a case of trilateral interactions', *Entomologia Experimentalis et Applicata*, 110(1), pp. 11–21.
Francis, F. *et al.* (2022) 'Complementary strategies for biological control of aphids and related virus transmission in sugar beet to replace neonicotinoids', *Agriculture*, 12(10), p. 1663.
Hardie, J. (2017) 'Life cycles and polyphenism', in *Aphids as crop pests.* 2nd ed. CABI, pp. 81–97.
Harmel, N. *et al.* (2008) 'Identification of aphid salivary proteins: A proteomic investigation of Myzus persicae', *Insect Molecular Biology*, 17(2), pp. 165–174. doi: 10.1111/j.1365-2583.2008.00790.x
Hemming, D. *et al.* (2022) 'Likelihood of extreme early flight of Myzus persicae (Hemiptera: Aphididae) across the UK', *Journal of Economic Entomology*, 115(5), pp. 1342–1349.
Hewer, A., Becker, A. and van Bel, A. J. E. (2011) 'An aphid's Odyssey–the cortical quest for the vascular bundle', *Journal of Experimental Biology*, 214(22), pp. 3868–3879.

Hogenhout, S. A. and Bos, J. I. B. (2011) 'Effector proteins that modulate plant–insect interactions', *Current Opinion in Plant Biology*, 14(4), pp. 422–428.

Holman, J. (2009) *Host plant catalog of aphids: palaearctic region*. Springer.

Jaouannet, M. *et al.* (2014) 'Plant immunity in plant–aphid interactions', *Frontiers in Plant Science*, 5, p. 663.

Jeon, H.-Y. *et al.* (2008) 'Economic injury level of Myzus persicae (Homoptera: Aphididae) at Chinese cabbage', *Korean Journal of Applied Entomology*, 47(4), pp. 407–411.

Kallure, G. S. *et al.* (2022) 'Characterized constituents of insect herbivore oral secretions and their influence on the regulation of plant defenses', *Phytochemistry*, 193, p. 113008.

Khan, M. A. *et al.* (2011) 'Response of Myzus persicae (Sulzer) to imidacloprid and thiamethoxam on susceptible and resistant potato varieties', *Sarhad Journal of Agriculture*, 27(2), pp. 263–269.

Kim, J. H. *et al.* (2008) 'Identification of indole glucosinolate breakdown products with antifeedant effects on Myzus persicae (green peach aphid)', *The Plant Journal*, 54(6), pp. 1015–1026.

Kindlmann, P. and Dixon, A. F. G. (1989) 'Developmental constraints in the evolution of reproductive strategies: telescoping of generations in parthenogenetic aphids', *Functional Ecology*, 3(5)pp. 531–537.

Li, Y. *et al.* (2016) 'Insecticide resistance monitoring and metabolic mechanism study of the green peach aphid, Myzus persicae (Sulzer) (Hemiptera: Aphididae), in Chongqing, China', *Pesticide Biochemistry and Physiology*, 132, pp. 21–28.

London, H. *et al.* (2020) 'The ability of the green peach aphid (Myzus persicae) to penetrate mesh crop covers used to protect potato crops against tomato potato psyllid (Bactericera cockerelli)', *PeerJ*, 8, p. e9317.

Machado-Assefh, C. R. *et al.* (2015) 'Disrupting Buchnera aphidicola, the endosymbiotic bacteria of Myzus persicae, delays host plant acceptance', *Arthropod-Plant Interactions*, 9(5), pp. 529–541.

Margaritopoulos, J. T. *et al.* (2002) 'Life cycle variation of Myzus persicae (Hemiptera: Aphididae) in Greece', *Bulletin of Entomological Research*, 92(4), pp. 309–319. doi: 10.1079/ber2002167

Margaritopoulos, J. T. *et al.* (2007) 'Insecticide resistance status of Myzus persicae (Hemiptera: Aphididae) populations from peach and tobacco in mainland Greece', *Pest Management Science: formerly Pesticide Science*, 63(8), pp. 821–829.

Margaritopoulos, J. T. *et al.* (2021) 'Long-term studies on the evolution of resistance of Myzus persicae (Hemiptera: Aphididae) to insecticides in Greece', *Bulletin of Entomological Research*, 111(1), pp. 1–16.

Mauck, K. E., De Moraes, C. M. and Mescher, M. C. (2010) 'Deceptive chemical signals induced by a plant virus attract insect vectors to inferior hosts', *Proceedings of the National Academy of Sciences*, 107(8), pp. 3600–3605.

Naga, K. C. *et al.* (2020) 'Transmission efficiency of Potato leaf roll virus (PLRV) by potato aphid Aulacorthum solani and green peach aphid Myzus persicae', *Indian Journal of Entomology*, 82(1), pp. 68–71.

Nalam, V., Louis, J. and Shah, J. (2019) 'Plant defense against aphids, the pest extraordinaire', *Plant Science*, 279, pp. 96–107.

Nampeera, E. L. (2022) *Management of green peach aphid, Myzus persicae in amaranth using host plant resistance and seed treatment*. JKUAT-CoANRE.

Nampeera, E. L. *et al.* (2023) 'Effects of seed treatments and storage duration on Myzus persicae (Hemiptera: Aphididae) and amaranth fresh leaf yield', *Environmental Entomology*, 52(3), pp. 360–370.

Nancarrow, N. *et al.* (2022) 'Symptomless turnip yellows virus infection causes grain yield loss in lentil and field pea: A three-year field study in south-eastern Australia', *Frontiers in Plant Science*, 13, p. 4756.

Nauen, R. and Denholm, I. (2005) 'Resistance of insect pests to neonicotinoid insecticides: current status and future prospects', *Archives of Insect Biochemistry and Physiology: Published in Collaboration with the Entomological Society of America*, 58(4), pp. 200–215.

Pareja, M. *et al.* (2012) 'Herbivory by a phloem-feeding insect inhibits floral volatile production', *PLoS One*, 7(2), p. e31971.

Park, K. C. and Hardie, J. (2002) 'Functional specialisation and polyphenism in aphid olfactory sensilla', *Journal of Insect Physiology*, 48(5), pp. 527–535.

Petek, M. *et al.* (2014) 'Potato virus Y infection hinders potato defence response and renders plants more vulnerable to Colorado potato beetle attack', *Molecular Ecology*, 23(21), pp. 5378–5391.

Pinheiro, P. F. *et al.* (2013) 'Insecticidal activity of citronella grass essential oil on Frankliniella schultzei and Myzus persicae', *Ciência e Agrotecnologia*, 37, pp. 138–144.

Pitino, M. and Hogenhout, S. A. (2013) 'Aphid protein effectors promote aphid colonization in a plant species-specific manner', *Molecular Plant–Microbe Interactions*, 26(1), pp. 130–139.

Pym, A. *et al.* (2022) 'Overexpression of UDP-glucuronosyltransferase and cytochrome P450 enzymes confers resistance to sulfoxaflor in field populations of the aphid, Myzus persicae', *Insect Biochemistry and Molecular Biology*, 143, p. 103743.

Qi, Y.-H. *et al.* (2021) 'Physical contact transmission of Cucumber green mottle mosaic virus by Myzus persicae', *PloS One*, 16(6), p. e0252856.

Schwartzberg, E. G. and Tumlinson, J. H. (2014) 'Aphid honeydew alters plant defence responses', *Functional Ecology*, 28(2), pp. 386–394.

Sharma, S., Sood, A. K. and Ghongade, D. S. (2022) 'Assessment of losses inflicted by the aphid, Myzus persicae (Sulzer) to sweet pepper under protected environment in north western Indian Himalayan region', *Phytoparasitica*, 50(1), pp. 51–62.

Shih, P.-Y., Sugio, A. and Simon, J.-C. (2023) 'Molecular mechanisms underlying host plant specificity in Aphids', *Annual Review of Entomology*, 68, pp. 431–450.

Sidauruk, L. and Sipayung, P. (2018) 'Population of Myzus persicae (Sulzer) and insect diversity on intercropping potatoes with other plants which planting at different time', in *IOP conference series: earth and environmental science*. IOP Publishing, p. 12018.

Silva-Sanzana, C. *et al.* (2022) 'Oligogalacturonides enhance resistance against aphids through pattern-triggered immunity and activation of salicylic acid signaling', *International Journal of Molecular Sciences*, 23(17), p. 9753.

Simon, J.-C. and Peccoud, J. (2018) 'Rapid evolution of aphid pests in agricultural environments', *Current Opinion in Insect Science*, 26, pp. 17–24.

Singh, B., Jasrotia, P. and Crespo-Herreraa, L. (2022) 'Breeding for aphid resistance in wheat: status and future prospects', in *New horizons in wheat and barley research*, Editors: Prem Lal Kashyap, Vikas Gupta, Om Prakash Gupta, R. Sendhil, K. Gopalareddy, Poonam Jasrotia, Gyanendra Pratap Singh, pp. 381–399, Springer Singapore.

Soderlund, D. M. and Bloomquist, J. R. (1990) 'Molecular mechanisms of insecticide resistance', in *Pesticide resistance in arthropods*. Springer, pp. 58–96.

Stevens, M. *et al.* (2008) *Turnip yellows virus (syn Beet western yellows virus): an emerging threat to European oilseed rape production*. HGCA Cape Town.

Tagu, D. *et al.* (2008) 'Early progress in aphid genomics and consequences for plant–aphid interactions studies', *Molecular Plant–Microbe Interactions*, 21(6), pp. 701–708.

Tamaki, G. (1975) 'Weeds in orchards as important alternate sources of green peach aphids in late spring', *Environmental Entomology*, 4(6), pp. 958–960.

Tamaki, G. and Fox, L. (1982) 'Weed species hosting viruliferous green peach aphids, vector of beet western yellows virus', *Environmental Entomology*, 11(1), pp. 115–117.

Troczka, B. J. *et al.* (2021) 'Molecular innovations underlying resistance to nicotine and neonicotinoids in the aphid Myzus persicae', *Pest Management Science*, 77(12), pp. 5311–5320.

Uriel, Y. (2020) 'Genetic flexibility in the green peach aphid: Defensive polyphenism in response to parasitoid pressure and the assembly of a new draft genome'. https://summit.sfu.ca/item/34294
van Emden, H. F. and Harrington, R. (2007) *Aphids as crop pests*. 1st edn. Edited by H. F. van Emden and R. Harrington. CABI.
van Emden, H. F. *et al.* (1969) 'The ecology of Myzus persicae', *Annual Review of Entomology*, 14(1), pp. 197–270.
Wakil, W., Brust, G. E. and Perring, T. D. (2017) *Sustainable management of arthropod pests of tomato*. Academic Press.
Westwood, J. H. *et al.* (2014) 'Interference with jasmonic acid-regulated gene expression is a general property of viral suppressors of RNA silencing but only partly explains virus-induced changes in plant–aphid interactions', *The Journal of General Virology*, 95(Pt 3), p. 733.
Will, T. and van Bel, A. J. E. (2006) 'Physical and chemical interactions between aphids and plants', *Journal of Experimental Botany*, 57(4), pp. 729–737.
Williams, I. S. and Dixon, A. F. G. (2007) 'Life cycles and polymorphism', in *Aphids as crop pests*, Editors: H. F. van Emden, R. Harrington, pp. 69–85, CABI.
Zhou, S. *et al.* (2015) 'Alteration of plant primary metabolism in response to insect herbivory', *Plant Physiology*, 169(3), pp. 1488–1498.
Zhuang, H. *et al.* (2018) 'Aphid (Myzus persicae) feeding on the parasitic plant dodder (Cuscuta australis) activates defense responses in both the parasite and soybean host', *New Phytologist*, 218(4), pp. 1586–1596.
Züst, T. and Agrawal, A. A. (2016) 'Mechanisms and evolution of plant resistance to aphids', *Nature Plants*, 2(1), pp. 1–9.

4 Current Management Strategies, Challenges, and Proposed Solutions

4.1 INTRODUCTION

The peach-potato aphid, *Myzus persicae*, is a ubiquitous pest that inflicts significant economic damage on a variety crops around the world. Its ability to reproduce rapidly and transmit plant viruses have made it a significant threat to agriculture, and, therefore, it has been a subject of intensive research. The management of *M. persicae* is a challenging task due to its high reproductive rate and the development of insecticide resistance. Scientists are working around the globe to develop new control strategies against destructive agricultural pests (Nauen *et al.*, 2015). Consequently, there are several pest management practices that have been in use for a considerable period, which fall under the following categories: chemical, biological and cultural methods (Barzman *et al.*, 2015). However, there are several challenges associated with these practices. Among these are the adverse effect of chemical control methods on non-target organisms, limitations with regard to biological control, a lack of resistant cultivars, and the underutilisation of cultural practices and defense elicitors. Overall, the effective management of *M. persicae* requires an integrated approach that takes into account the challenges associated with current management practices.

Therefore, it is highly necessary to develop new alternatives to combat or reduce its population to below the level at which economic injury might occur (Dara, 2019). To avoid the side effects of the new development management strategies it is essentially important to consider negative effects on environment and non-target organisms. Recent studies have been reported that several non-chemical options, such as biological control and plant resistance, and cultural practices, training, workshops and awareness program about integrated pest management approaches can provide effective and environmentally friendly ways to manage *M. persicae* populations (Dardouri *et al.*, 2017; Kadzamira *et al.*, 2022). However, the success of these approaches depends on various factors, including the complexity of the pest–crop–environment interaction, and hence requires a multidisciplinary approach to find the most appropriate solution for each situation.

The aim of this chapter is to review the current management strategies for *M. persicae* and examine the challenges facing researchers and farmers in their control. This chapter also provides a comprehensive overview of the current state of knowledge on *M. persicae* management practices and highlights future research directions in the efforts towards controlling this pest.

DOI: 10.1201/9781003400974-4

4.2 CURRENT MANAGEMENT PRACTICES

The *M. persicae* is a prevalent pest for various crops, which causes significant economic losses due to yield reduction and virus transmission. Therefore, effective management practices are essential to control this insect and prevent further damage. In this section, we will review the current management practices used to control *M. persicae* and discuss their effectiveness and sustainability.

4.2.1 Cultural Practices

Cultural practices are the first line of defense against *M. persicae* (Grechi, Sauge, *et al.*, 2008), as they deal with habitat management, variety selection, and the timing of planting (Altieri and Nicholls, 2004; Bajwa and Kogan, 2004). Cultural practices that can be used to reduce the impact of *M. persicae* include: (1) *Early planting*: This practice may reduce the severity of *M. persicae* infestations, as earlier planted crops may pass through the growth stages before peak aphid populations emerge. (2) *Row spacing*: Widening of the row spaces can create an environment that is not conducive to the development of *M. persicae*. This is because wider rows allow for better air movement and light penetration, thereby reducing humidity and temperature, and making it harder for aphids to establish and reproduce (Hooks and Fereres, 2006). (3) *Soil management*: Regular soil tillage practices to bury crop residue can reduce the proliferation of aphid populations. This action has the added benefit of decreasing the risk of overwintering of the aphids (Perring *et al.*, 2018). (4) *Variety selection*: Relatively resistant crop varieties may be used as a passive control measure against *M. persicae* (Ali, Sobhy and Bruce, 2022). The identification of aphid-resistant varieties can be achieved through plant breeding, genetic modification, or selection of the least preferred host.

4.2.2 Mechanical Control

Mechanical control encompasses physical techniques that aim to reduce the aphid population (Barzman *et al.*, 2015; Adhikari, 2022). Among the mechanical measures are: (1) *Combing*: Combing of the crop foliage can physically remove aphids from the plants (Hooks and Fereres, 2006). Studies have shown that mechanical control is effective when carried out repeatedly and systematically. It has also been shown to be a more sustainable option than chemical application. (2) *Trapping*: Yellow sticky traps can be used for population monitoring and to prevent winged aphids from colonizing new plants. These traps are effective but require frequent monitoring and replacement (Murtaza *et al.*, 2019). (3) *High-pressure water sprays*: High-pressure water sprays directly targeted at aphids can result in significant mortality, as the pressure causes physical damage to the aphids (Nawaz, Ali and Abbas, 2016).

4.2.3 Biological Control

Biological control is an alternative to chemical control approaches which involves a number of living organisms belonging to the different phylum and kingdoms such

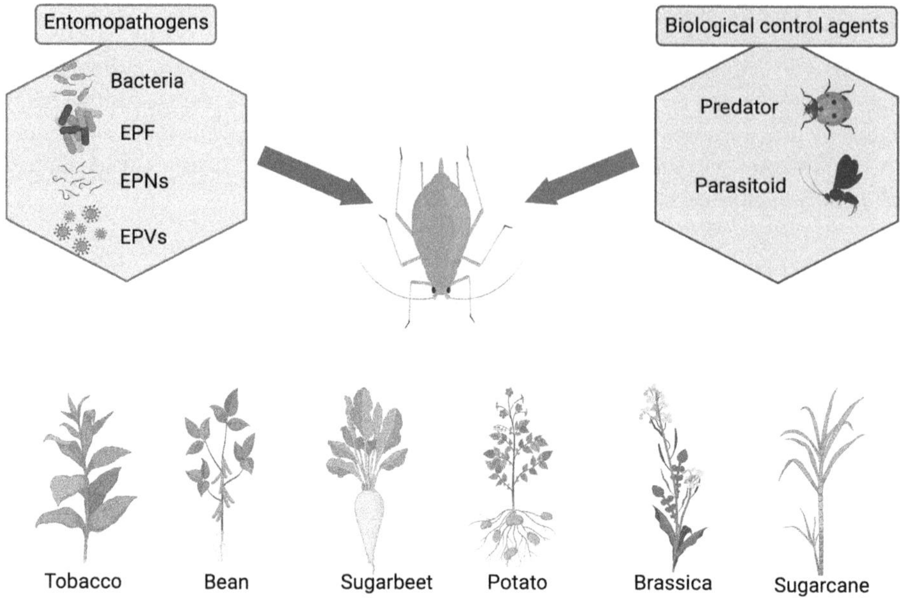

FIGURE 4.1 Biological control options to manage the peach-potato aphid *Myzus persicae*, EPF: Entomopathogenic fungi; EPNs: Entomopathogenic nematodes; EPVs: Entomopathogenic virus.

as microbes (fungi, bacteria and viruses), nematodes and insects (parasitoids and predators belonging to the order Hymenoptera and Coleoptera) (Sanda and Sunusi, 2014) (Figure 4.1).

Biocontrol agents are widely used in pest management (Barratt *et al.*, 2018). Extensive research has been done to test the efficacy of such biological control agents on *M. persicae* both in the laboratory and in the field (Andorno and López, 2014). Approximately 200 biocontrol agents belonging to different families (*Coccinellidae, Cantharidae, Syrphidae, Anthocoridae, Pentatomidae, Aphelinidae, Braconidae, Phytoseiidae*) have been identified as potential insects for the population management of *M. persicae* (van Emden *et al.*, 1969; Mohammed and Hatcher, 2017). The host plant plays a central role in implementing the biocontrol approach of naturally occurring biological control agents. The effectiveness of biocontrol agents depends on the species and the physiological status of host plants. Plants release volatile compounds, which can be used as a cue by pollinators and other pests' natural enemies to locate the host plants (Xu and Turlings, 2018). In particular, the approximate numbers of predators (150), parasitoids (50) and entomopathogens (40) have been reported in earlier studies.

Entomopathogens are naturally occurring microorganisms, such as bacteria, fungi, viruses, and nematodes, which can infect and kill insect pests (Paliwal *et al.*, 2022). One such pest is *M. persicae*, or the peach-potato aphid, which has been identified as a major global agricultural pest that feeds on a range of crops. Several entomopathogens have been reported as being capable of controlling this pest. For instance, when bacterial insecticides has been applied a high reduction in the population of

M. persicae was recorded; in one study a diet containing bacterial suspension showed 100% mortality of adult and nymphs (López-Isasmendi *et al.*, 2019), in another a 57% population reduction on cauliflower was observed (Akbar, Rana and Perveen, 2014), and in a third *Capsicum annum*, *A. thaliana* and *Beta vulgaris* showed population reductions of 68%, 57% and 69%, respectively, on the application of *P. fluorescens* (PpR24) (Paliwal *et al.*, 2022). Entomopathogenic fungi (EPF) *Beauveria bassiana* and *Metarhizium anisopliae* have both been reported as a potential control method for more than 300 species of arthropods, including aphids (*M. persicae*) (Khachatourians and Qazi, 2008; de Barros, Fronza and Bertholdo-Vargas, 2015; Yun *et al.*, 2017; Bamisile *et al.*, 2021).

Additionally, entomopathogenic viruses (EPVs) also have negative effects on the peach-potato aphid: for instance, a reduction in the growth of *M. persicae* has been recorded on tobacco plants infested with potato virus-Y (PVY) (Ren *et al.*, 2015), as has an increased susceptibility to parasitoids (Mauck, De Moraes, and Mescher, 2015). Furthermore, a few genera of nematodes are important entomopathogens that affect the performance of the attacking herbivore on the host plant by inducing plant defense (activates polyphenol oxidase and guaiacol peroxidase activity in roots and at the same suppressing the same activity in above ground tissue) (Johnson and Rasmann, 2015; Kamali *et al.*, 2022). However, limited work has been done to test their efficacy against aphids, particularly *M. persicae*; for example, *Steinernema carpocapsae* showed low infectivity when tested on *M. persicae* (Park et al., 2013).

4.2.4 Chemical Control

Chemical control involves the use of pesticides to control *M. persicae* populations (Dewar and Denholm, 2017). While chemical control is often necessary in cases of severe infestation or to prevent the spread of viral diseases, it is essential to use these products judiciously, considering their potential adverse impact on non-target organisms, such as beneficial insects, and the environment. The following are among the chemical control measures that can be employed: (1) *Insecticides*: Insecticides are a quick and effective solution to aphid infestations (Morita *et al.*, 2007). However, the extensive use of insecticides can lead to the development of resistance in the target pest population. It is therefore necessary to rotate between different classes of insecticides and reduce the reliance on broad-spectrum sprays. (2) *Systemic insecticides*: Systemic insecticides can be applied to crops, where they are taken up by the plant's vascular system, making these chemicals toxic to the aphids that feed on the plant (Halder, Kodandaram and Rai, 2011). Systemic insecticides are effective, but their use is discouraged with some crops since residual toxicity can harm non-target organisms. (3) *Seed treatments*: Seed treatments have the advantage of providing early control of aphids, especially in seedling crops (Seagraves and Lundgren, 2012). They also have the added benefit of efficient use of the pesticide since only a small amount of the product is required.

4.3 CHALLENGES WITH CURRENT MANAGEMENT PRACTICES

Managing *M. persicae* populations requires a multifaceted approach which combines cultural and biological practices in a balanced manner. The important aspect of the management approach is the establishment of an integrated approach, by using non-chemical methods. Despite an extensive research undertaken on the development of management strategies against *M. persicae*, agricultural production is still suffering from this pest. It could be due to non-target and/or side effects of management strategies, a lack of awareness, side effects and the underutilization of non-chemical control approaches. After reviewing the published literature on control strategies, we have concluded several points which seem major obstacles with current management practices.

4.3.1 Lack of Resistant Cultivars

Insects have also adapted to overcome these defenses, creating a constant arms race between plants and their herbivorous attackers (Gatehouse, 2002). Studies observe that *M. persicae* has shown the ability to overcome plant resistance mechanisms, thereby neutralizing the benefits of genetically modified or resistant plant varieties (Elzinga, De Vos and Jander, 2014). Plant defense against herbivores is an ongoing challenge which requires constant innovation and management. As the natural world continues to evolve, it is important for researchers and farmers to remain vigilant and adapt their strategies to keep up with new insect threats. By working together, we can develop more effective and sustainable approaches to protect our crops and natural habitats from insect herbivores.

The use of aphid-resistant cultivars with integrated pest management approaches appears to be an excellent control option for *M. persicae* (Tetreault *et al.*, 2019). One of the significant differences between aphid-resistant and aphid-susceptible lines of the same crop is the length of probing time taking by the aphid in reaching up to the phloem (Dreyer and Campbell, 1987). On resistant lines, aphids take a long time to reach the phloem and they are unable to feed successfully (Klingler *et al.*, 1998, 2005). In contrast, aphids take a relatively short probing time on susceptible lines compared to resistant lines (Greenslade *et al.*, 2016). In addition to this, aphids also have difficulty in feeding on resistant lines in locating the phloem during probing (Ahmed, 2002).

4.3.2 Adverse Effect on Non-Target Organisms

Literature on the side effects of insecticides on arthropods reported that more than 400 agricultural chemicals has adverse effects on natural enemies, which vary from sublethal to lethal (Theiling and Croft, 1988). Excessive use of synthetic pesticides is responsible not only for resistance in insects but also has deleterious effects on non-target organisms such as pollinators, predators, and parasitoids (Stanley, Preetha and Stanley, 2016; Serrão *et al.*, 2022). Parasitoids, which are of great economic significance in pest management and play a vital role in managing the pest population, are badly affected by the excessive use of these synthetic chemicals (Kampfraath *et al.*, 2017). Studies have shown that inappropriate use and application timing of insecticides affect the parasitoid population, emergence rate, longevity, reproductive rate and parasitic behavior (Ulber, Klukowski and Williams, 2010; El-Wakeil *et al.*, 2013). Synthetic chemicals such as fenitrothion and Spinosad (organophosphates) affect the longevity and reproductive capacity of egg parasitoids *Trissolcus grandis* and *Trichogramma* sp. (Shoeb, 2010; El-Wakeil *et al.*, 2013). Similarly, deltamethrin (pyrethroids) treatment of crops has a negative impact on the emergence and parasitic behavior of *T. grandis* and *D. rapae* respectively (Desneux *et al.*, 2005).

4.3.3 Elicitation of Plant Defense: An Underutilized Tool

Despite excellent results, the use of natural defense compounds as a plant defense elicitor (under field conditions) is still in its early stages (Jamiołkowska, 2020; Ali *et al.*, 2021). The majority of these studies are limited to the laboratory and are still waiting for applications in the field, for instance; *cis*-Jasmone (CJ) is a natural plant-derived compound, which has been tested on several economically important crops (potato, brassica, wheat, tomato) against various sucking pests, including *M. persicae* (Birkett *et al.*, 2000; Ali *et al.*, 2021). Many laboratory studies have been performed which investigate the role of CJ in inducing resistance to pests; almost all these studies proved that CJ induces defense against insect pests (Birkett *et al.*, 2000; Bayram and Tonğa, 2018; Ali *et al.*, 2021). A recent study on the CJ treatment of five brassica genotypes showed a high level of induced resistance against *M. persicae* (Ali *et al.*, 2021). Similarly, an extensive literature is available that reported the high potential of a range of plant extracts in managing the pests. These included: olive leaf extract caused 100% *M. persicae* (at 10% concentration); *Xanthium strumarium, Tanacetum parthenium* and *Hypericum calycinum* plant extracts showed 89%, 88% and 57% nymphal mortality, respectively, when *M. persicae* was exposed to treated leaf disk (Hamouda *et al.*, 2015; Erdoğan and Yıldırım, 2016; Pavela, 2018). Similarly, extracts of *Ricinus communis, Solanum nigrum* and *Nerium oleander* were found to be highly toxic and showed a high level of *M. persicae* mortality in a toxicity bioassay (Madanat, Al Antary and Zarqa, 2016).

4.3.4 Underutilization of Cultural Practices

Cultural methods are normally used to create unfavorable conditions for pests, which is one of the main goals of pest management approaches (Abate, van Huis and Ampofo, 2000). Cultural practices such as crop rotation and sanitation, intercropping, the destruction of plant debris, the avoidance of adjacent planting of crops and trap crops can all play a critical role in pest management (Brader, 1979; Smit and Matengo, 1995). For instance, winter peach pruning is a practice (particularly in fruit orchards to adjust the plant load) which is considered a means of controlling vegetative growth of plants that indirectly affect the performance of pest, i.e. *M. persicae* (Grechi, Sauge, *et al.*, 2008). It has been reported that *M. persicae* was predominantly found on mature leaves, so winter pruning significantly affects the population of this pest (Grechi, Sauge, *et al.*, 2008; Grechi *et al.*, 2012). Similarly, intercropping companion plants with horticultural crops can be an effective method for pest management; for instance, planting *Tagetes patula* as a companion plant with capsicum annuum significantly impacts the population of *M. persicae* (Dardouri *et al.*, 2017, 2019). However, due to the wide range of host plants, cultural practices may be a complicated and daunting task to control *M. persicae* (Capinera, 2001).

4.3.5 Limitations with Biological Control

Biological control using predators and parasitoids is an effective strategy to manage aphids, but it can be challenging to establish and maintain a stable population of natural enemies in the field because it takes a long time to develop and implement a successful biological control program (Barratt *et al.*, 2018). Cost can also be a factor, including the development, testing, and release of the biological control agent (Askary, 2015). Furthermore, methods are often not practical on large commercial farms, and their efficacy can vary depending on the environment.

4.3.6 Regulatory Restrictions

Some insecticides used for *M. persicae* management are being phased out as a result of environmental concerns (Dewar, 2017; Hauer *et al.*, 2017). This makes it important to develop alternative management strategies that are environmentally sustainable. Furthermore, insecticide regulations and registration process are becoming increasingly stringent, making it challenging to find effective and safe solutions to manage insect herbivores (Birch, 2011; Sobhy *et al.*, 2020).

4.4 PROPOSED SOLUTIONS

4.4.1 Applying Ecological Principles

Ecological principles such as increasing biodiversity and implementing intercropping techniques have been found to be effective in managing this pest. Increasing biodiversity is an effective method of reducing pest populations as it provides an abundance of natural enemies that feed on pests (Tiwari, Sharma and Wratten, 2020). This helps to maintain balance in the ecosystem and reduces the need for chemical pesticides that are harmful to the environment and cause resistance in pest populations. A study by Sidauruk and Sipayung (2018) found that planting diverse plant species in sunflower fields increased the abundance of natural predators of herbivores, resulting in a significant reduction in aphid populations (Middleton and MacRae, 2021). Intercropping companion plants is another effective way of managing *M. persicae* (Dardouri *et al.*, 2019; Tiwari, Sharma and Wratten, 2020). Companion planting involves planting two or more species together which have a mutually beneficial relationship, such as halting the growth of pest populations (Dardouri *et al.*, 2017).

Research has shown that intercropping garlic with tobacco and celery, maize, sunflower with potato significantly reduced the population of *M. persicae* and increased the yield of crops (Lai *et al.*, 2011; Sidauruk and Sipayung, 2018). A recent study performed on the effect of companion plants against *M. persicae* showed that *Tagetes patula, Basel* released VOCs that directly and indirectly affect the performance and behavior of *M. persicae* when used as companion plant with *Capsicum annuum* in the field (Dardouri *et al.*, 2019). Conclusively, ecological principles such as increasing biodiversity and implementing intercropping techniques have been found to be effective in managing *M. persicae*. These methods are sustainable, affordable, and do not require the use of harmful chemical pesticides, making them extremely beneficial to the environment and the economy.

4.4.2 Traps and Lures

Use of traps and lures is another important method that is used for mating disruption and play a critical role in integrated pest management (IPM) to control *M. persicae* populations (Grechi, Sauphanor, *et al.*, 2008; Javed *et al.*, 2016). Traps are devices that are used to attract and capture insects. In the case of *M. persicae*, yellow sticky traps can be used to trap adult aphids (Saljoqi, Khan and Rahman, 2009). Yellow is a known attractant to many insects and has been shown to be effective in trapping. The use of traps can help to monitor populations and provide an early indication of an infestation. By trapping adults, the possibility of egg laying and further infestations is reduced, leading to better pest management control.

On the other hand, lures are devices that are used to attract insects through the release of chemicals or other substances that mimic natural attractants. Lures can be used to attract and trap *M. persicae*, helping to reduce the application of pesticides. A well-targeted lure can attract adult aphids from their host plants, making them easier to control. Furthermore, the use of lures and traps aligns with a more holistic approach to pest management (Kaur *et al.*, 2021), as the chemicals used for these purposes do not have any side effects on non-target organisms and are also eco-friendly. Overall, traps and lures are useful components of an integrated pest management plan, as they can help reduce pest populations and minimize the exposure of plants to harmful chemicals.

4.4.3 Augmentative Release of Biocontrol Agents

The augmentative release of biocontrol agents against aphids is one of the most effective and sustainable methods to control these pests (Zhou *et al.*, 2014). Such release involves releasing natural enemies, such as parasitoids, predators, and entomopathogenic fungi or bacteria, to control aphid populations. It reduces the need to use chemical insecticides and thereby does not have negative effects on non-target organisms and the environment (Van Lenteren, 2000). In addition, biocontrol agents are specific to a particular pest and do not harm beneficial insects such as bees, butterflies, and ladybirds. Biocontrol agents can establish themselves in an ecosystem and provide for the long-term control of aphids. This can help reduce the future risk of aphid outbreaks. This promotes healthy and sustainable agricultural practices that can benefit both farmers and the environment.

4.4.4 Genetic Engineering

Genetic engineering can potentially be used against aphid pests by introducing genetic modifications into crops that make them more resistant to aphid infestations (Yu *et al.*, 2012; Murtaza *et al.*, 2022). One approach would be to introduce genes that produce proteins toxic to aphids (Romeis, Meissle and Bigler, 2006). This would allow crops to produce their own insecticide and reduce the need for chemical pesticides. Another approach could involve introducing genes that produce plant natural defense compounds, such as terpenes, phenolics, and glucosinolates, that effectively deter aphids from feeding on the plants. Overall, the development of genetically engineered crops that are resistant to pests and diseases can reduce the need for pesticides.

4.4.5 Host Plant Resistance

Host plant resistance is a key component of integrated pest management strategies for the management of insect herbivores (Stout, 2014). Host plant resistance play a vital role in reducing the infestation of herbivores in two broad ways i.e. direct defense and indirect defense. Host plant direct defense involves the use of plant traits that deter or reduce insect pests, thereby reducing damage to the plant and minimizing the need for chemical pesticides (Mouden and Leiss, 2021). For instance, research has shown that host plant resistance can be highly effective in managing *M. persicae* populations. Similarly, a study by Ali *et al.* (2022) found that potato cultivars with high levels of resistance to *M. persicae* had significantly lower aphid populations and damage to the plants compared to susceptible cultivars. Another study by Hao *et al.* (2020) showed that *M. persicae* feeding behaviour varied according to the susceptibility of the brassica cultivars. Studies have also found that other plant traits, such as leaf pubescence and trichome density, can affect herbivores' behavior and development (Levin, 1973; War *et al.*, 2012; Javed *et al.*, 2021).

Indirect host plant resistance involves the production of allelochemicals, which are compounds that are toxic or repellent to insect pests and act as an attractant for their natural enemies (predators and parasitoids). For example, some Brassica species produce glucosinolates, which are toxic to aphid and other insect herbivores (Newton, Bullock and Hodgson, 2009). However, it is important to note that host plant resistance may be ineffective against all insect pests, and that it is affected by environmental conditions and genetic variability within pest populations. In addition, some insect pests can overcome host plant resistance through the rapid evolution of resistance mechanisms. Therefore, host plant resistance should be used in combination with other management strategies, such as biological control and cultural practices, to achieve more effective and sustainable pest management.

4.4.6 Induced Resistance

Induced resistance is a plant defense mechanism that involves the activation of defense genes in response to biotic or abiotic stress. Various biotic and abiotic agents or products have been applied to prime plants to enhance their induced resistance against pest attacks. Recent studies have shown that induced defense can be a potential tool to induce plant defense against *M. persicae* (Ali *et al.*, 2021). Several studies have shown the effectiveness of induced resistance against *M. persicae* (Basit *et al.*, 2019; Javed and Qiu, 2020; Ali *et al.*, 2021). One study (Song and Ryu, 2013) found that 3-pentanol and 2-butanone to cucumber plants induced the production of volatile organic compounds (VOCs) that increase the abundance of natural enemies of *M. persicae*. This enhanced emission of VOCs is probably due to the activation of defense-related pathways (oxylipin) in response to 3-pentanol and 2-butanone treatment.

In addition to biotic agents, abiotic factors, such as water stress, can also alter the defense responses of plants against herbivores. For instance, plants grown under limited water conditions were found to be highly susceptible to *M. persicae* and showed a low level of secondary metabolites compared to plants that were grown under proper (plenty of) water supply (Khan, Ulrichs and Mewis, 2010). One study, Quandahor *et al.* (2021), found that exposing potato cultivar (cultivar) to mild drought stress increased their resistance to *M. persicae* by inducing the expression of defense genes. Overall, induced resistance has been shown to be an effective approach to control *M. persicae* in various crop species. The use of biotic and abiotic agents/products to prime plants can enhance their resistance to this pest, reducing the need for synthetic pesticides and promoting sustainable pest management practices.

4.4.7 Precision Agriculture

Precision agriculture is the use of technology to optimize crop production by monitoring crop yield, nutrient level, weather conditions, and other factors. Here are some examples and references of precision agriculture for pest management: Drones equipped with multispectral cameras can be used to detect aphid infestations in crops. The cameras can detect changes in the reflectance of light that are associated with aphid feeding damage. By monitoring crops from the air, farmers can quickly identify areas of infestation and take appropriate action. Robotic control of aphids refers to the use of robots or autonomous machines to control aphids, which are a common pest in many agricultural settings (Lacotte *et al.*, 2022). The goal of robotic control is to reduce the use of harmful pesticides and other chemicals that can have negative impacts on the environment and human health. One approach to robotic control involves the use of autonomous drones that can identify and target aphids with precision. These drones can use advanced sensors to detect the presence of aphids on crops, and then use robotic arms or other mechanisms to deliver targeted doses of insecticides or other treatments to the affected areas.

Another approach involves the use of robotic "AgBots" that can navigate through fields and orchards, scanning for insect herbivore infestations and applying treatments as needed (Redhead *et al.*, 2015; Bechar and Vigneault, 2016). These machines can be programmed to follow specific patterns or routes through a field, and can also be equipped with sensors that can monitor the environment and detect changes in temperature, humidity, and other factors that may affect aphid populations. Overall, the goal of robotic control of aphids is to reduce the damage caused by these pests, while minimizing the negative impacts of chemicals and other treatments on the environment, human health, and other beneficial organisms.

4.5 CONCLUSION

In conclusion, *M. persicae* significantly impacts global food security, and the effective management of these pests remains a daunting challenge. To date, various strategies, such as chemical control, biological control, cultural practices, defense elicitors, and regulatory restrictions, have been implemented to manage the pest population. However, the effectiveness of these methods has been limited due to some challenges, including adverse effects on non-target organisms, limitations with biological control, plant defense being overcome, a lack of resistant cultivars, the underutilization of cultural practices and defense elicitors, and regulatory restrictions. Therefore, there is a need to develop new holistic management strategies to combat the pest while reducing the limitations of the current strategies. This approach can significantly contribute to a sustainable pest management system, ensuring that food remains accessible, safe, and affordable for everyone.

REFERENCES

Abate, T., van Huis, A. and Ampofo, J. K. O. (2000) 'Pest management strategies in traditional agriculture: an African perspective', *Annual Review of Entomology*, 45(1), pp. 631–659.

Adhikari, U. (2022) 'Insect pest management: mechanical and physical techniques', *Reviews in Food and Agriculture (RFNA)*, 3(1), pp. 48–53.

Ahmed, K. A. (2002) *An integrated approach to pest management in field pea, Pisum sativum (L.), with emphasis on pea aphid, Acyrthosiphon pisum (Harris).* University of the Free State.

Ahmed, Q. *et al.* (2022) 'Response of aphid parasitoids to volatile organic compounds from undamaged and infested brassica oleracea with Myzus persicae', *Molecules*, 27(5), p. 1522.

Akbar, M. F., Rana, H. U. and Perveen, F. (2014) 'Management of cauliflower aphid (Myzus persicae (Sulzer) Aphididae: Hemiptera) through environment friendly bioinsecticides', *Pakistan Entomologist*, 36(1), pp. 25–30.

Ali, J., Sobhy, I. S. and Bruce, T. J. A. (2022) 'Wild potato ancestors as potential sources of resistance to the aphid Myzus persicae', *Pest Management Science*, 78(9), pp. 3931–3938.

Ali, J. *et al.* (2021) 'Effects of cis-jasmone treatment of brassicas on interactions with myzus persicae aphids and their parasitoid diaeretiella rapae', *Frontiers in Plant Science*, 12. doi: 10.3389/fpls.2021.711896

Altieri, M. and Nicholls, C. (2004) *Biodiversity and pest management in agroecosystems.* CRC Press.

Andorno, A. V. and López, S. N. (2014) 'Biological control of Myzus persicae (Hemiptera: Aphididae) through banker plant system in protected crops', *Biological Control*, 78, pp. 9–14.

Askary, T. H. (2015) 'Limitations, research needs and future prospects in the biological control of phytonematodes', in *Biocontrol agents of phytonematodes*, CABI Publishing, pp. 446–454.

Badiali, C. *et al.* (2018) 'Chitosan oligosaccharides affect xanthone and VOC biosynthesis in Hypericum perforatum root cultures and enhance the antifungal activity of root extracts', *Plant Cell Reports*, 37(11), pp. 1471–1484. doi: 10.1007/s00299-018-2317-2

Bajwa, W. I. and Kogan, M. (2004) 'Cultural practices: springboard to IPM', in *Integrated pest management: Potential, constraints and challenges*, Editors: O. Koul, G. S. Dhaliwal, G. W. Cuperus, CABI Publishing, pp. 21–38.

Bamisile, B. S. *et al.* (2021) 'Model application of entomopathogenic fungi as alternatives to chemical pesticides: Prospects, challenges, and insights for next-generation sustainable agriculture', *Frontiers in Plant Science*, 12, p. 741804.

Barratt, B. I. P. *et al.* (2018) 'The status of biological control and recommendations for improving uptake for the future', *BioControl*, 63(1), pp. 155–167.

Barzman, M. *et al.* (2015) 'Eight principles of integrated pest management', *Agronomy for Sustainable Development*, 35(4), pp. 1199–1215.

Basit, A. *et al.* (2019) 'Molecular and functional characterization of elicitor PeBC1 extracted from Botrytis cinerea involved in the induction of resistance against green peach aphid (Myzus persicae) in common beans (Phaseolus vulgaris L.)', *Insects*, 10(2), p. 35.

Bayram, A. and Tonğa, A. (2018) 'cis-Jasmone treatments affect pests and beneficial insects of wheat (*Triticum aestivum* L.): the influence of doses and plant growth stages', *Crop Protection*, 105, pp. 70–79.

Bechar, A. and Vigneault, C. (2016) 'Agricultural robots for field operations: concepts and components', *Biosystems Engineering*, 149, pp. 94–111.

Birch, E., Nicholas, A., Begg, G. S., and Geoffrey, R. (2011) 'How agro-ecological research helps t o address food security issues under new IPM and pesticide reduction policies fo r global crop production systems', *Journal of Experimental Botany*, 62(10), pp. 3251–3261.

Birkett, M. A. *et al.* (2000) 'New roles for cis-jasmone as an insect semiochemical and in plant defense', *Proceedings of the National Academy of Sciences of the United States of America*, 97(16), pp. 9329–9334.

Brader, L. (1979) 'Integrated pest control in the developing world', *Annual Review of Entomology*, 24(1), pp. 225–254.

Capinera, J. L. (2001) *Green peach aphid, myzus persicae (sulzer)(insecta: hemiptera: aphididae).* University of Florida Cooperative Extension Service, Institute of Food and Agricultural Sciences, EDIS.

Casteel, C. L. *et al.* (2014) 'The NIa-Pro protein of Turnip mosaic virus improves growth and reproduction of the aphid vector, Myzus persicae (green peach aphid)', *The Plant Journal*, 77(4), pp. 653–663.

Chakraborty, M. *et al.* (2020) 'Mechanism of plant growth promotion and disease suppression by chitosan biopolymer', *Agriculture*, 10(12), p. 624.

Chandler, D. *et al.* (2011) 'The development, regulation and use of biopesticides for integrated pest management', *Philosophical Transactions of the Royal Society B: Biological Sciences*, 366(1573), pp. 1987–1998.

Dara, S. K. (2019) 'The new integrated pest management paradigm for the modern age', *Journal of Integrated Pest Management*, 10(1), p. 12.

Dardouri, T. *et al.* (2017) 'How French marigold (*Tagetes patula* L.) volatiles can affect the performance of green peach aphid', *IOBC-WPRS Bull*, 123, pp. 71–78.

Dardouri, T. *et al.* (2019) 'Repellence of Myzus persicae (Sulzer): evidence of two modes of action of volatiles from selected living aromatic plants', *Pest Management Science*, 75(6), pp. 1571–1584.

De, A. *et al.* (2014) *Targeted delivery of pesticides using biodegradable polymeric nanoparticles.* Springer.

de Barros, N. M., Fronza, E. and Bertholdo-Vargas, L. R. (2015) 'Use of fungi as biopesticides', in *Biopesticides handbook*, Springer Nature Switzerland, pp. 247–281.

Der Geest, V. (2000) 'Can plants use entomopathogens as bodyguards?', *Ecology Letters*, 3(3), pp. 228–235.

Desneux, N. *et al.* (2005) 'Diaeretiella rapae limits Myzus persicae populations after applications of deltamethrin in oilseed rape', *Journal of Economic Entomology*, 98(1), pp. 9–17.

Dewar, A. M. (2017) 'The adverse impact of the neonicotinoid seed treatment ban on crop protection in oilseed rape in the United Kingdom', *Pest Management Science*, 73(7), pp. 1305–1309.

Dewar, A. M. and Denholm, I. (2017) 'Chemical control', in *Aphids as crop pests*, Editors: H. F. van Emden, R. Harrington, pp. 398–425.

Dewhirst, S. Y., Pickett, J. A. and Hardie, J. (2010) 'Aphid pheromones', *Vitamins & Hormones*, 83, pp. 551–574.

Diabate, S. *et al.* (2021) 'Push-pull strategy combined with net houses for controlling cowpea insect pests and enhancing crop yields', *Crop Protection*, 141, p. 105480.

Dreyer, D. L. and Campbell, B. C. (1987) 'Chemical basis of host-plant resistance to aphids', *Plant, Cell & Environment*, 10(5), pp. 353–361.

El-Wakeil, N. *et al.* (2013) 'Side effects of insecticides on natural enemies and possibility of their integration in plant protection strategies', in *Agricultural and biological sciences "insecticides—development of safer and more effective technologies"*, Editor: Stanislav Trdan, Intech, Rijeka, Croatia, pp. 1–54.

Elzinga, D. A., De Vos, M. and Jander, G. (2014) 'Suppression of plant defenses by a Myzus persicae (green peach aphid) salivary effector protein', *Molecular Plant–Microbe Interactions*, 27(7), pp. 747–756.

Erdoğan, P. and Yıldırım, A. (2016) 'Insecticidal activity of three different plant extracts on the green peach aphid [(Myzus persicae Sulzer)(Hemiptera: Aphididae)]', *Journal of the Entomological Research Society*, 18(1), pp. 27–35.

Gatehouse, J. A. (2002) 'Plant resistance towards insect herbivores: a dynamic interaction', *New Phytologist*, 156(2), pp. 145–169.

Grechi, I., Sauge, M., *et al.* (2008) 'How does winter pruning affect peach tree–Myzus persicae interactions?', *Entomologia Experimentalis et Applicata*, 128(3), pp. 369–379.

Grechi, I., Sauphanor, B., *et al.* (2008) 'Evaluation of integrated management scenarios of the peach tree-Myzus persicae system using a crop-pest model', in *Dear Colleagues It Was with the Great Pleasure We Welcomed You to the 7th IOBC Conference on Integrated Fruit Production at Avignon in 2008. The Programme for This Meeting Was Exceptionally Rich and Diverse and Contained Many Interesting and Exciting Cont*, p. 312.

Grechi, I. *et al.* (2012) 'Designing integrated management scenarios using simulation-based and multi-objective optimization: Application to the peach tree–Myzus persicae aphid system', *Ecological Modelling*, 246, pp. 47–59.

Greenslade, A. F. C. *et al.* (2016) 'Triticum monococcum lines with distinct metabolic phenotypes and phloem-based partial resistance to the bird cherry–oat aphid Rhopalosiphum padi', *Annals of Applied Biology*, 168(3), pp. 435–449.

Halder, J., Kodandaram, M. H. and Rai, A. B. (2011) 'Differential responses of major vegetable aphids to newer insecticide molecules', *Vegetable Science*, 38(2), pp. 191–193.

Hamouda, A. Ben *et al.* (2015) 'Studies on insecticidal and deterrent effects of olive leaf extracts on Myzus persicae and Phthorimaea operculella', *Journal of Entomology and Zoology Studies*, 3(6), pp. 294–297.

Hao, Z.-P. *et al.* (2020) 'Possible effects of leaf tissue characteristics of oilseed rape Brassica napus on probing and feeding behaviors of cabbage aphids *Brevicoryne brassicae*', *Arthropod-Plant Interactions*, 14, pp. 733–744.

Hauer, M. *et al.* (2017) 'Neonicotinoids in sugar beet cultivation in Central and Northern Europe: Efficacy and environmental impact of neonicotinoid seed treatments and alternative measures', *Crop Protection*, 93, pp. 132–142.

Helms, A. M. *et al.* (2019) 'Chemical cues linked to risk: cues from below-ground natural enemies enhance plant defences and influence herbivore behaviour and performance', *Functional Ecology*, 33(5), pp. 798–808.

Hooks, C. R. R. and Fereres, A. (2006) 'Protecting crops from non-persistently aphid-transmitted viruses: a review on the use of barrier plants as a management tool', *Virus Research*, 120(1–2), pp. 1–16.

Jamiołkowska, A. (2020) 'Natural compounds as elicitors of plant resistance against diseases and new biocontrol strategies', *Agronomy*, 10(2), p. 173.

Javed, K. and Qiu, D. (2020) 'Protein elicitor PeBL1 of Brevibacillus laterosporus enhances resistance against Myzus persicae in tomato', *Pathogens*, 9(1), p. 57.

Javed, K. *et al.* (2016) 'Field evaluation of different insecticides against green peach aphid Myzus persicae (Sulzer)', *Journal of Entomology and Zoology Studies*, 4, pp. 657–662.

Javed, K. *et al.* (2021) 'PeaT1 and PeBC1 microbial protein elicitors enhanced resistance against Myzus persicae Sulzer in chili *Capsicum annum* L.', *Microorganisms*, 9(11), p. 2197.

Jiménez-Reyes, M. F. *et al.* (2019) 'Natural compounds: a sustainable alternative to the phytopathogens control', *Journal of the Chilean Chemical Society*, 64(2), pp. 4459–4465.

Johnson, S. N. and Rasmann, S. (2015) 'Root-feeding insects and their interactions with organisms in the rhizosphere', *Annual Review of Entomology*, 60(1), pp. 517–535.

Kadzamira, M. A. T. J. *et al.* (2022) 'A non-linear approach to the establishment of local biological control agent production units: a case study of fall armyworm in Bangladesh', *CABI Agriculture and Bioscience*, 3(1), p. 48.

Kamali, S. *et al.* (2022) 'Beneficial worm allies warn plants of parasite attack below-ground and reduce above-ground herbivore preference and performance', *Molecular Ecology*, 31(2), pp. 691–712.

Kampfraath, A. A. *et al.* (2017) 'Pesticide stress on plants negatively affects parasitoid fitness through a bypass of their phytophage hosts', *Ecotoxicology*, 26, pp. 383–395.

Kaur, K. *et al.* (2021) 'Nanomaze lure: Pheromone sandwich in graphene oxide interlayers for sustainable targeted pest control', *ACS Applied Materials & Interfaces*, 13(41), pp. 48349–48357.

Khachatourians, G. G. and Qazi, S. S. (2008) 'Entomopathogenic fungi: biochemistry and molecular biology', in *Human and animal relationships*, Editors: Axel A. Brakhage, Peter F. Zipfel, Springer, pp. 33–61.

Khan, M. A. M., Ulrichs, C. and Mewis, I. (2010) 'Influence of water stress on the glucosinolate profile of Brassica oleracea var. italica and the performance of Brevicoryne brassicae and Myzus persicae', *Entomologia Experimentalis et Applicata*, 137(3), pp. 229–236.

Kim, S. *et al.* (2018) 'Evaluating natural compounds as potential insecticides against three economically important pests, Bemisia tabaci (Hemiptera: Aleyrodidae), Frankliniella occidentalis (Thysanoptera: Thripidae), and Myzus persicae (Hemiptera: Aphididae), on greenhouse swee', *Applied Biological Chemistry*, 61(3), pp. 313–323.

Klingler, J. *et al.* (1998) 'Phloem specific aphid resistance in Cucumis melo line AR 5: effects on feeding behaviour and performance of Aphis gossypii', *Entomologia Experimentalis et Applicata*, 86(1), pp. 79–88.

Klingler, J. *et al.* (2005) 'Aphid resistance in Medicago truncatula involves antixenosis and phloem-specific, inducible antibiosis, and maps to a single locus flanked by NBS-LRR resistance gene analogs', *Plant Physiology*, 137(4), pp. 1445–1455.

Lacotte, V. *et al.* (2022) 'Pesticide-free robotic control of aphids as crop pests', *AgriEngineering*, 4(4), pp. 903–921.

Lai, R. *et al.* (2011) 'Response of green peach aphids and other arthropods to garlic intercropped with tobacco', *Agronomy Journal*, 103(3), pp. 856–863.

Levin, D. A. (1973) 'The role of trichomes in plant defense', *The Quarterly Review of Biology*, 48(1, Part 1), pp. 3–15.

Li, Y. *et al.* (2020) 'Modulation of above-belowground plant-herbivore interactions by entomopathogenic nematodes', *Applied Soil Ecology*, 148, p. 103479.

Liu, J. *et al.* (2019) 'Preference of the aphid Myzus persicae (Hemiptera: Aphididae) for tobacco plants at specific stages of potato virus Y infection', *Archives of Virology*, 164(6), pp. 1567–1573.

López-Isasmendi, G. *et al.* (2019) 'Aphicidal activity of Bacillus amyloliquefaciens strains in the peach-potato aphid (Myzus persicae)', *Microbiological Research*, 226, pp. 41–47.

Madanat, H. M., Al Antary, T. M. and Zarqa, M. (2016) 'Toxicity of six ethanol plant extracts against the green peach aphid Myzus persicae Sulzer (Homoptera: Aphididae)', *Fresenius Environmental Bulletin*, 25(3), pp. 706–718.

Mauck, K. E., De Moraes, C. M. and Mescher, M. C. (2015) 'Infection of host plants by Cucumber mosaic virus increases the susceptibility of Myzus persicae aphids to the parasitoid Aphidius colemani', *Scientific Reports*, 5(1), pp. 1–9.

Middleton, E. G. and MacRae, I. V. (2021) 'Wildflower plantings in commercial agroecosystems promote generalist predators of Colorado potato beetle', *Biological Control*, 152, p. 104463.

Mohammed, A. A. and Hatcher, P. E. (2017) 'Combining entomopathogenic fungi and parasitoids to control the green peach aphid Myzus persicae', *Biological Control*, 110, pp. 44–55.

Morita, M. *et al.* (2007) 'Flonicamid, a novel insecticide with a rapid inhibitory effect on aphid feeding', *Pest Management Science: Formerly Pesticide Science*, 63(10), pp. 969–973.

Mouden, S. and Leiss, K. A. (2021) 'Host plant resistance to thrips (Thysanoptera: Thripidae)–current state of art and future research avenues', *Current Opinion in Insect Science*, 45, pp. 28–34.

Murtaza, G. *et al.* (2019) 'Effectiveness of different traps for monitoring sucking and chewing insect pests of crops', *Egyptian Academic Journal of Biological Sciences A, Entomology*, 12(6), pp. 15–21.

Murtaza, S. *et al.* (2022) 'Silencing a myzus persicae macrophage inhibitory factor by plant-mediated RNAi induces enhanced aphid mortality coupled with boosted RNAi efficacy in transgenic potato lines', *Molecular Biotechnology*, 64(10), pp. 1152–1163.

Nauen, R. *et al.* (2015) 'Flupyradifurone: a brief profile of a new butenolide insecticide', *Pest Management Science*, 71(6), pp. 850–862.

Nawaz, M., Ali, S. and Abbas, Q. (2016) 'High pressure water spray technique for controlling mustard aphid (Lipaphis Erysimi) on Brassica crop', *American-Eurasian Journal of Agricultural & Environmental Sciences*, 16(2), pp. 224–228.

Newton, E. L., Bullock, J. M. and Hodgson, D. J. (2009) 'Glucosinolate polymorphism in wild cabbage (Brassica oleracea) influences the structure of herbivore communities', *Oecologia*, 160, pp. 63–76.

Paliwal, D. *et al.* (2022) 'Identification of novel aphid-killing bacteria to protect plants', *Microbial Biotechnology*, 15(4), pp. 1203–1220.

Paprocka, M. *et al.* (2018) 'Novel hydroxy- and epoxy-cis-jasmone and dihydrojasmone derivatives affect the foraging activity of the peach potato aphid Myzus persicae (Sulzer) (Homoptera: Aphididae)', *Molecules*, 23(9), p. 2362. doi: 10.3390/molecules23092362

Park, H. W. *et al.* (2013) 'Infectivity of entomopathogenic nematode Steinernema carpocapsae Pocheon strain (Nematoda: Steinernematidae) on the green peach aphid Myzus persicae (Hemiptera: Aphididae) and its parasitoids', *Biocontrol Science and Technology*, 23(6), pp. 637–645.

Pavela, R. (2018) 'Essential oils from Foeniculum vulgare Miller as a safe environmental insecticide against the aphid Myzus persicae Sulzer', *Environmental Science and Pollution Research*, 25, pp. 10904–10910.

Perring, T. M. *et al.* (2018) 'Aphids: biology, ecology, and management', in *Sustainable management of arthropod pests of tomato*, Editors: Waqas Wakil, Gerald E. Brust, Thomas M. Perring, Elsevier, pp. 15–48.

Quandahor, P. *et al.* (2021) 'Comparison of root tolerance to drought and aphid (Myzus persicae Sulzer) resistance among different potato (*Solanum tuberosum* L.) cultivars', *Scientific Reports*, 11(1), pp. 1–13.

Ragaei, M. and Sabry, A. H. (2014) 'Nanotechnology for insect pest control', *International Journal of Science, Environment and Technology*, 3(2), pp. 528–545.

Redhead, F. *et al.* (2015) 'Bringing the farmer perspective to agricultural robots', in *Proceedings of the 33rd Annual ACM Conference Extended Abstracts on Human Factors in Computing Systems*, pp. 1067–1072.

Ren, G. *et al.* (2015) 'Potato virus Y-infected tobacco affects the growth, reproduction, and feeding behavior of a vector aphid, Myzus persicae (Hemiptera: Aphididae)', *Applied Entomology and Zoology*, 50(2), pp. 239–243.

Romeis, J., Meissle, M. and Bigler, F. (2006) 'Transgenic crops expressing Bacillus thuringiensis toxins and biological control', *Nature Biotechnology*, 24(1), pp. 63–71.

Saljoqi, A. U. R., Khan, K. and Rahman, S. U. (2009) 'Integrated management of potato-peach aphid, Myzus persicae (Sulzer)', *Sarhad Journal of Agriculture*, 25(4), pp. 573–580.

Sanda, N. B. and Sunusi, M. (2014) 'Fundamentals of biological control of pests', *International Journal of Clinical & Biological Sciences*, 1(6), pp. 1–11.

Saroj, A. *et al.* (2020) 'Phytochemicals of plant-derived essential oils: A novel green approach against pests', in *Natural remedies for pest, disease and weed control*, Editors: Chukwuebuka Egbuna, Barbara Sawicka, Elsevier, pp. 65–79.

Sathiyabama, M., Akila, G. and Charles, R. E. (2014) 'Chitosan-induced defence responses in tomato plants against early blight disease caused by Alternaria solani (Ellis and Martin) Sorauer', *Archives of Phytopathology and Plant Protection*, 47(16), pp. 1963–1973.

Seagraves, M. P. and Lundgren, J. G. (2012) 'Effects of neonicitinoid seed treatments on soybean aphid and its natural enemies', *Journal of Pest Science*, 85, pp. 125–132.

Senthil-Nathan, S. (2020) 'A review of resistance mechanisms of synthetic insecticides and botanicals, phytochemicals, and essential oils as alternative larvicidal agents against mosquitoes', *Frontiers in Physiology*, 10, p. 1591.

Serrão, J. E. *et al.* (2022) 'Side-effects of pesticides on non-target insects in agriculture: A mini-review', *The Science of Nature*, 109(2), p. 17.

Shoeb, M. A. (2010) 'Effect of some insecticides on the immature stages of the egg parasitoid Trichogramma evanescens West.(Hym., Trichogrammatidae)', *Egyptian Academic Journal of Biological Sciences. A, Entomology*, 3(1), pp. 31–38.

Sidauruk, L. and Sipayung, P. (2018) 'Population of Myzus persicae (Sulzer) and insect diversity on intercropping potatoes with other plants which planting at different time', in *IOP conference series: earth and environmental science*. IOP Publishing, p. 12018.

Smit, N. and Matengo, L. O. (1995) 'Farmers' cultural practices and their effects on pest control in sweetpotato in South Nyanza, Kenya', *International Journal of Pest Management*, 41(1), pp. 2–7.

Sobhy, I. S. *et al.* (2020) 'Sensing the danger signals: cis-jasmone reduces aphid performance on potato and modulates the magnitude of released volatiles', *Frontiers in Ecology and Evolution*, 7(January), pp. 1–14. doi: 10.3389/fevo.2019.00499

Song, G. C. and Ryu, C.-M. (2013) 'Two volatile organic compounds trigger plant self-defense against a bacterial pathogen and a sucking insect in cucumber under open field conditions', *International Journal of Molecular Sciences*, 14(5), pp. 9803–9819.

Souto, A. L. *et al.* (2021) 'Plant-derived pesticides as an alternative to pest management and sustainable agricultural production: Prospects, applications and challenges', *Molecules*, 26(16), p. 4835.

Stanley, Johnson, Preetha, G. and Stanley, J (2016) *Pesticide toxicity to non-target organisms*. Springer, pp. 351–410.

Stout, M. J. (2014) 'Host-plant resistance in pest management', in *Integrated pest management*, Editors: Dharam P. Abrol, Elsevier, pp. 1–21.

Tetreault, H. M. *et al.* (2019) 'Global responses of resistant and susceptible sorghum (Sorghum bicolor) to sugarcane aphid (Melanaphis sacchari)', *Frontiers in Plant Science*, 10, p. 145.

Theiling, K. M. and Croft, B. A. (1988) 'Pesticide side-effects on arthropod natural enemies: a database summary', *Agriculture, Ecosystems & Environment*, 21(3–4), pp. 191–218.

Tiwari, S., Sharma, S. and Wratten, S. D. (2020) 'Flowering alyssum (Lobularia maritima) promote arthropod diversity and biological control of Myzus persicae', *Journal of Asia-Pacific Entomology*, 23(3), pp. 634–640.

Ulber, B., Klukowski, Z. and Williams, I. H. (2010) 'Impact of insecticides on parasitoids of oilseed rape pests', in *Biocontrol-based integrated management of oilseed rape pests*, Editor: Ingrid H. Williams, Springer Dordrecht, pp. 337–355.

van Emden, H. F. *et al.* (1969) 'The ecology of Myzus persicae', *Annual Review of Entomology*, 14(1), pp. 197–270.

Van Lenteren, J. C. (2000) 'Success in biological control of arthropods by augmentation of natural enemies', in *Biological control: measures of success*, Editors: Geoff Gurr, Steve Wratten. Springer, pp. 77–103.

van Munster, M. *et al.* (2003) 'Characterization of a new densovirus infecting the green peach aphid Myzus persicae', *Journal of Invertebrate Pathology*, 84(1), pp. 6–14.

Van Munster, M. *et al.* (2005) 'Can plants use an entomopathogenic virus as a defense against herbivores?', *Oecologia*, 143(3), pp. 396–401.

Wang, D. *et al.* (2022) 'Nano-enabled pesticides for sustainable agriculture and global food security', *Nature Nanotechnology*, 17(4), pp. 347–360.

War, A. R. *et al.* (2012) 'Mechanisms of plant defense against insect herbivores', *Plant Signaling and Behavior*, 7(10). doi: 10.4161/psb.21663

Weintraub, P. G. (2009) 'Physical control: an important tool in pest management programs', in *Biorational control of arthropod pests: application and resistance management*, Editors: Isaac Ishaaya, A. Rami Horowitz, pp. 317–324.

Xu, H. and Turlings, T. C. J. (2018) 'Plant volatiles as mate-finding cues for insects', *Trends in Plant Science*, 23(2), pp. 100–111.

Yu, X. *et al.* (2012) '(E)-β-Farnesene synthase genes affect aphid (Myzus persicae) infestation in tobacco (Nicotiana tabacum)', *Functional & Integrative Genomics*, 12, pp. 207–213.

Yun, H.-G. *et al.* (2017) 'Entomopathogenic fungi as dual control agents against both the pest Myzus persicae and phytopathogen Botrytis cinerea', *Mycobiology*, 45(3), pp. 192–198.

Zhou, H. *et al.* (2014) 'Biological control of insect pests in apple orchards in China', *Biological Control*, 68, pp. 47–56.

Zhu, J. *et al.* (2006) 'Sex pheromone of the soybean aphid, Aphis glycines Matsumura, and its potential use in semiochemical-based control', *Environmental Entomology*, 35(2), pp. 249–257.

5 Successful Examples

5.1 INTRODUCTION

The peach-potato aphid *Myzus persicae* is a major pest which affects a wide range of crops worldwide. Traditional methods of pest control, such as chemical insecticides, have been shown to have negative impacts on both the environment and non-target organisms, and can also lead to the development of resistance in the targeted pest populations. As a result, alternative methods of pest control have been developed that are effective against *M. persicae* while minimizing negative impacts on the environment and non-target organisms. In recent years, there has been a growing concern about the environmental impact and sustainability of chemical insecticides in controlling *M. persicae*. As a result, there is a pressing need to develop alternative and ecologically-friendly management strategies against this pest. Previous studies have reported several successful control measures tested against peach-potato aphid using non-toxic chemical strategies, such as the use of biocontrol agents, resistant cultivars, and entomopathogens.

Furthermore, the induction of plant defense by the application of natural compounds, such as plant extracts or elicitors, is another effective method of controlling *M. persicae*. These compounds activate the natural defense mechanisms of plants, making them less susceptible to *M. persicae* infestation. Similarly, priming plants, which involves the application of low doses of an elicitor, has also been shown to be an effective method of controlling insect herbivores (Birkett *et al.*, 2000; Dardouri *et al.*, 2019; J. Ali *et al.*, 2021a; Eizi, 2021). By priming plants, they are able to mount an enhanced defense response against *M. persicae*, reducing the damage caused by infestations. Overall, these methods of pest control provide effective alternatives to chemical insecticides in controlling *M. persicae*, while minimizing negative impacts on the environment and non-target organisms. Their use can help to ensure sustainable pest management practices in agriculture.

This chapter reviews successful examples of such management strategies, including crop rotation, intercropping, inducing plant defense, producing resistant plant cultivars, and developing entomopathogens. The chapter describes the scientific basis and practical considerations of each approach, highlights their benefits and challenges, and discusses their potential integration for effective and sustainable *M. persicae* management. These examples provide insights and inspirations for researchers, farmers, and policymakers who are interested in promoting sustainable agriculture and pest management.

5.2 RESISTANT PLANT CULTIVARS

The domestication of crop plants often focuses on increasing the yield from crop plants and this could compromise plant defense, especially if such strategies have a yield penalty and are therefore selected against (Rodriguez-Saona *et al.*, 2011).

DOI: 10.1201/9781003400974-5

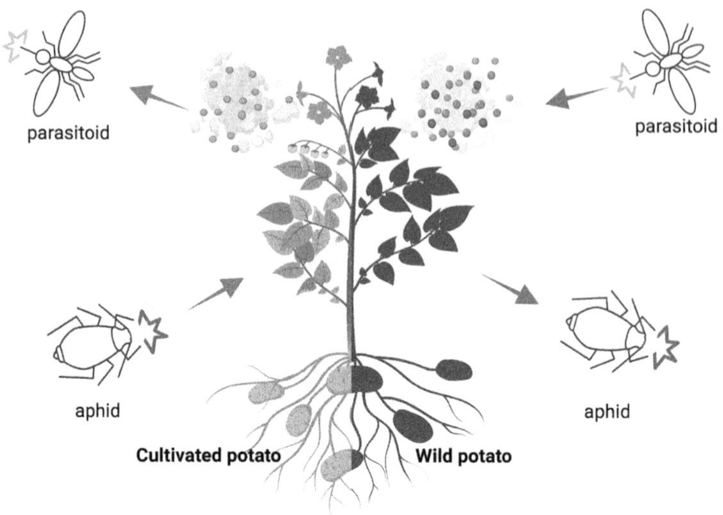

FIGURE 5.1 A summary of *M. persicae* and parasitoid *D. rapae* performance and behavioral responses on wild potato *S. stoloniferum* accessions (Ali *et al.*, 2022). A series of behavioral and performance bioassays were conducted. In aphid performance clip-cage bioassay, wild potato lines showed high adult aphid mortality and low larviposition. Similarly, a volatile analysis of wild and cultivated potato lines showed a high emission of VOCs in blends released by wild potato lines. These volatile samples were used for olfactometer bioassay, in which aphid *M. persicae* showed a negative response to volatile samples collected from wild potato lines. In contrast, parasitoid *D. rapae* showed a positive response to volatiles samples released by wild potato lines.

Previous studies have shown that breeding programs (with the primary goal of achieving high yield) have a negative impact on host plant resistance that make crops more susceptible to pest due to the manipulation of its genetics to produce high yields (Gepts, 2002; Mishra *et al.*, 2015). To test this hypothesis, we investigated the performance of *M. persicae* on three wild potato *Solanum stoloniferum* and one cultivated potato *Solanum tubersoum* cv. Desiree lines (Ali, Sobhy and Bruce, 2022). Our results showed that all wild potato lines were highly resistant against *M. persicae* and cause high adult mortality in cage bioassay. Similarly, olfactometer studies showed that adult *M. persicae* spent less time in olfactometer arm treated with plant volatile samples collected from wild potato lines. In contrast, volatile blends released by wild potato lines were highly attractive to *D. rapae*, one of the main natural enemies of *M. persicae*. Chapter 6 will discuss more about this study in detail (Figure 5.1).

5.3 INDUCED PLANT DEFENSE

Being in continuous co-evolution with insects, plants have evolved a broad armory of effective and sophisticated defense mechanisms that enable plants to defend themselves against these invading insects accordingly before they cause significant damage (Howe and Jander, 2008). Such an array of defensive strategies involves the production of a plethora of different chemical compounds covering numerous classes of toxic and/or repellent secondary metabolites, digestibility-reducing proteins, and antinutritive enzymes that either kill or severely affect the herbivores' growth (Mithöfer and Boland, 2012). As such, plants tackle the herbivores directly by negatively influencing their preference (host plant selection, oviposition, feeding behavior) or performance (growth rate, development, reproductive success) resulting in an improved plant fitness. Examples of defensive secondary metabolites comprise alkaloids, cardenolides, glycosides, glucosinolates, flavonoids and phenolic acids, coumarins and furanocoumarins, protease inhibitors, terpenoids and tannins (Després, David and Gallet, 2007).

Some of these chemical defenses are constitutive, but the majority is induced soon after the invasion of herbivores (Mithöfer and Boland, 2012). The cost of induced defense responses in plants is much lower than that of constitutive resistance (Lu *et al.*, 2015). Given that many defense chemicals are high in terms of energy cost and have nutrient requirements associated with their production (Cipollini and Heil, 2010), plants thus only produce these chemicals once they discern any stimuli of invading insects, this phenomenon is known as induced plant defense (Karban and Myers, 1989). In this regard, plants also defend themselves indirectly against herbivores by releasing a complex bouquet of volatiles that recruit the natural enemies (parasitoid and predators) and/or providing reward (extra floral nectar) and shelter to enhance the foraging success of natural enemies (Heil, 2008). Furthermore, plant volatiles can also affect the herbivores themselves by repelling further colonization (De Moraes, Mescher and Tumlinson, 2001).

Interestingly, inducible defenses can be also activated using particular bioactive chemical compounds that are called elicitors or inducers of plant defense (Holopainen *et al.*, 2009). These findings have promoted the notion that plant-inducible defenses can be further enhanced or manipulated by switching on the defense genes responsible for key defense compounds (Kappers *et al.*, 2005). The genetic manipulation of induced plant defense has also been used successfully to alter the emissions of several volatile compounds and thereby enhance the attraction of natural enemies (Brillada *et al.*, 2013). This approach may offer future solutions; in most systems, however, we still miss fundamental knowledge on which key attractants should be targeted to achieve this approach (Sobhy *et al.*, 2014) and there could be costs if induced defenses are expressed constitutively.

Other options, such as the phenotypic manipulation of plant defense via application of chemical elicitors, has received considerable research attention to combat insect pests, particularly in the context of indirect plant defense and tritrophic interaction (J. Ali *et al.*, 2021a). During the last two decades, an accumulated body of investigations has been carried out to exploit the potential of using defense inducers to enhance plant immunity in general (Bektas and Eulgem, 2015). Nevertheless, while major advances have now been made in the application of plant inducers that activate resistance against plant pathogens (Gozzo and Faoro, 2013), most of the inducers promoting resistance to herbivores are still at the experimental level. Sustainable agriculture encourages reductions in the use of toxic chemical pesticides.

An extensive literature is available where natural compounds have been found effective in inducing plant defense. These include compounds such as: *cis*-Jasmone (CJ), β-ocimene, benzothiadiazole and methyl jasmonate chitosan, which have shown an

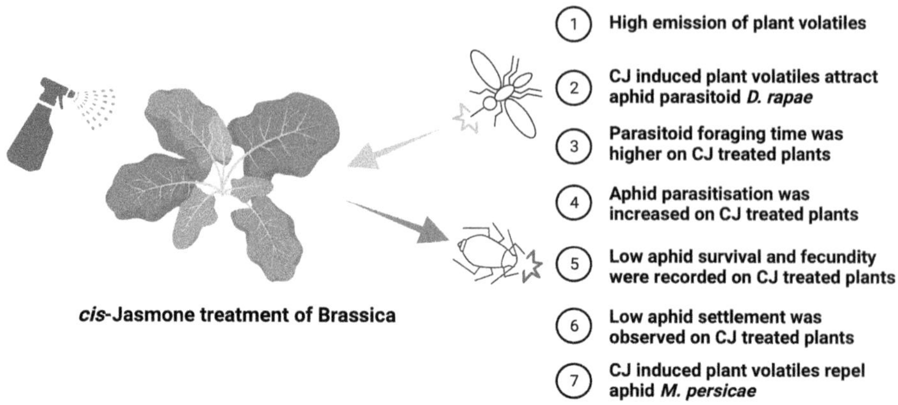

FIGURE 5.2 A summary of induced resistance and insects (*M. persicae* and *D. rapae*) responses to CJ-treated brassicas (Ali *et al.*, 2021b). Above study was conducted to investigate the effect of CJ on two brassica species (*Brassica rapa* and *Brassica napus*). A series of performance and behavioral bioassays were performed i.e., **(1)** In headspace sampling, CJ-treated plants showed a significant high qualitative and quantitative difference in volatile emission compared to untreated plant, **(2)** In olfactometer bioassay, parasitoid spent more time in the olfactometer arm treated with CJ treated plants' volatile, **(3)** In foraging bioassay, parasitoid spent more time on CJ treated plants, **(4)** In parasitism bioassay, high number of mummies were recorded on CJ-treated plants, **(5)** In aphid performance cage bioassay, high aphid mortality and low fecundity were recorded, **(6)** In aphid settlement bioassay, CJ-treated plants showed low aphid settlement compared to control, and **(7)** In olfactometer bioassay, *M. persicae* spent less time in the olfactometer arm treated with CJ-treated plants' volatiles.

elevated level of defense in plants (brassica, potato, tomato) (Boughton, Hoover and Felton, 2006; Kang *et al.*, 2018; J. Ali *et al.*, 2021a). In particular, the exogenous application of MeJA induces the formation of defense enzymes and a reduction in pupal/larval weights, performance, population densities and feeding behavior (Erbilgin *et al.*, 2006; Ma *et al.*, 2013; Cao, Wang and Liu, 2014; Xiao *et al.*, 2019; Wei *et al.*, 2021). CJ is another well-known example of a plant defense elicitor, which has been tested on several important crops including brassica, tomato, wheat, maize, sweet pepper against sucking pests (Bruce *et al.*, 2003; Matthes *et al.*, 2011; Oluwafemi *et al.*, 2013; Disi *et al.*, 2017; Sobhy *et al.*, 2017; J. Ali *et al.*, 2021a). A recent study on the exogenous application of MeSA on brassica (*Brassica rapa*) showed *M. persicae* performance was significantly reduced on MeSA treated plants compared to untreated (Ali *et al.*, 2023). Similarly, a detailed study conducted by Ali *et al.* (2021a) showed that CJ treatment of Brassica plants makes the plants less suitable for aphid *M. persicae* and more attractive to parasitoid *D. rapae*. Chapter 7 will discuss more about this study in details (Figure 5.2).

5.4 INTERCROPPING

Another important aspect of management strategies for *M. persicae* is intercropping; there are various studies against different herbivores that support the idea of planting different crops in the same field (Ben Issa, Gautier and Gomez, 2017; Ameline *et al.*, 2022). Recent studies performed on the effect of the companion plant against *M. persicae* showed that *T. patula, Basel* released VOCs that directly and indirectly affect the performance and behavior of *M. persicae* when used as companion plant with *C. annuum* in the field (Dardouri *et al.*, 2019). The volatile analysis of blend released by companion plants showed the presence of several compounds such as (*E*)-β-farnesene, β-linalool, caryophyllene, and pinene that have repellent effect on several herbivores including *M. persicae* (Oluwafemi *et al.*, 2013; Dardouri *et al.*, 2017, 2019).

An accumulating body of evidence reported the economic loss caused by *M. persicae* to a wide range of crop plants and several studies have been performed to develop the control strategies to control this pest. However, most of the studies are focusing on a single control measure either biological or chemical control, such as: effect of the host plant (Ali *et al.*, 2021b; Ali, Sobhy and Bruce, 2022); predators (Foglar *et al.*, 1990; Tiwari *et al.*, 2020); entomopathogens fungi (EPFs) (Lee *et al.*, 2015; Jaber and Araj, 2018; Singh and Joshi, 2020); entomopathogenic nematodes (EPNs) (Park *et al.*, 2013); entomopathogenic viruses (EPVs) (Ren *et al.*, 2015; Liu *et al.*, 2019) and push-pull (Cook, Khan and Pickett, 2007). Only a few pioneering studies have been undertaken to combine the biocontrol agents together, such as three species of EPF along with 1% azadirachtin (Lee *et al.*, 2015; Singh and Joshi, 2020); biocontrol recruitment via host plant-induced defense (Ali *et al.*, 2021b). In addition, the use of entomopathogens, botanical and chemical pesticides would more likely boost the control option (Barzman *et al.*, 2015). Furthermore, the use of natural compounds would be an additional protection for crops that will directly affect the pest population by producing defensive compounds and recruiting natural enemies by emitting induced plant volatile organic compounds (VOCs) (Ali *et al.*, 2021b). Results obtained from previous studies show that there is a potential to develop an IPM program based on multiple, complementary components (Naranjo, 2001; Jacobsen, Zidack and Larson, 2004; Cook, Khan and Pickett, 2007; Ali *et al.*, 2021b).

5.5 A PROPOSED IPM FOR *MYZUS PERSICAE*

An IPM pyramid provides detailed information about the actions required to control pests starting from the bottom (prevention), then progressing towards the top (chemical control) if the prevention and biocontrol strategies fail to suppress the population below the economic threshold (Lewis *et al.*, 1997; Stenberg, 2017). Agronomic practices has been extensively reviewed by Sunderland and Samu, (2000). Incorporating these practices would be a promising tool in controlling peach-potato aphid (Barzman *et al.*, 2015). Furthermore, it will also be helpful in reducing the use of chemical pesticides and enhanced biological control (Whipps and Lumsden, 2001; Symondson, Sunderland and Greenstone, 2002; Prakash and Rao, 2018). To synergise the biological control, use of conventional aphicide in integrated pest management (IPM) is also valid while adhering to IPM guidelines regarding chemical pesticides (Deguine *et al.*, 2021). To boost the IPM strategies, use of biocontrol agents in combination benefits the overall management strategies (Barzman *et al.*, 2015; Stenberg, 2017).

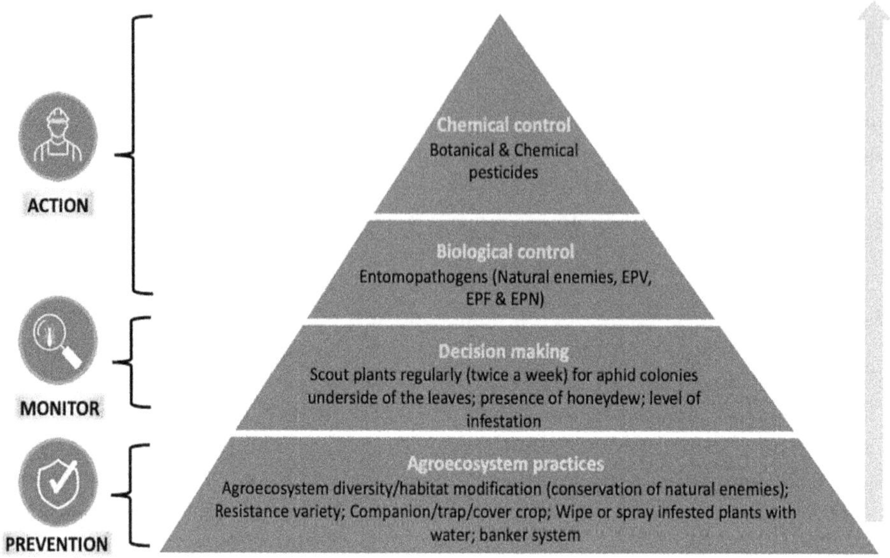

FIGURE 5.3 A proposed IPM pyramid for the control of peach-potato aphid, *Myzus persicae*. **(EPVs: entomopathogenic virus: EPF: entomopathogenic fungi; EPNs: entomopathogenic nematodes.)**

Furthermore, a wide range of entomopathogens have been employed in crop pest management that have the potential to control the pest and disease considerably (Ekesi, Dimbi and Maniania, 2007; Skinner, Parker and Kim, 2014). It would be better to use these entomopathogens as a preventive treatment particularly in early plant stages (Tomilova *et al.*, 2021). We have suggested a possible IPM pyramid accumulating all these essential component of IPM in Figure 5.3.

5.6 CONCLUSION

There are several obstacles that hinder the utilization of non-chemical pest control methods, including lack of awareness, financial instability, shortage of trained staff, and poor technical advancement (Rijal *et al.*, 2018). In developing countries, people are either unaware of integrated pest management approaches, or lack of trained staffs for successful implementation of control strategies (Rijal *et al.*, 2018). For example, a survey conducted in Chitwan, Nepal, revealed that only 17% of farmers received one short-term training on integrated pest management. Financial instability and poor technical advancement are also among the significant factors that impede the adoption of such potential approaches (Bottrell and Schoenly, 2018). In contrast, farmers show a high preference to chemical pesticides due to factors such as the falling prices of generic pesticides, their easy availability and readiness for application, and quick response (they are less time-consuming) (Denkyirah *et al.*, 2016; Abubakar *et al.*, 2020; Haggblade, Diarra and Traoré, 2022). According to a survey conducted in Chitwan, Nepal, 80% of the farmers showed a preference for chemical pesticides, even though 90% of them were aware of their side effects on the environment and non-target organisms (Rijal *et al.*, 2018).

In conclusion, below are several recommendations that may be helpful for future work:

- Resistant cultivars-based agriculture should be explored further to make the crop more resistant and less palatable to *M. persicae* (Ali, Sobhy and Bruce, 2022).
- As recommended by Ali *et al.* (2021b), the application of natural plant defense elicitors on the crop needs to be explored in the field.
- There is a need for more studies to investigate the effect of entomopathogens where the target pest is the peach-potato aphid.
- A combination of control strategies, such as resistant plant varieties, natural plant defense elicitors, entomopathogens, biocontrol agents, and biopesticides identified, need to be tested as part of integrated pest management rather than the application of these measures at individual level.
- More training and workshops are required to educate and train farmers about IPM-based approaches.

(Gao *et al.*, 2017; Dutta, Bhattacharyya and Kumari, 2023)

REFERENCES

Abubakar, Y., Tijjani, H., Egbuna, C., Adetunji, C.O., Kala, S., Kryeziu, T.L., Ifemeje, J.C. and Patrick-Iwuanyanwu, K.C., 2020. 'Pesticides, history, and classification', in *Natural remedies for pest, disease and weed control*. Academic Press, pp. 29–42.

Ali, J., Sobhy, I. S. and Bruce, T. J. A. (2022) 'Wild potato ancestors as potential sources of resistance to the aphid Myzus persicae', *Pest Management Science*, 78(9), pp. 3931–3938.

Ali, J. *et al.* (2021a) 'Effects of cis-jasmone treatment of brassicas on interactions with myzus persicae aphids and their parasitoid diaeretiella rapae', *Frontiers in Plant Science*, 12. doi: 10.3389/fpls.2021.711896

Ali, J. *et al.* (2023) 'Exogenous application of methyl salicylate induces defence in brassica against peach potato aphid myzus persicae', *Plants*, 12(9), p. 1770.

Ali, M. Y. *et al.* (2021b) 'Host-plant Variations affect the biotic potential, survival, and population projection of Myzus persicae (Hemiptera: Aphididae)', *Insects*, 12(5), p. 375.

Ameline, A. *et al.* (2022) 'Geranium macrorrhizum, a potential novel companion plant affecting preference and performance of Myzus persicae on sweet pepper', *Journal of Pest Science*, 96(2), pp. 671–682.

Barzman, M. *et al.* (2015) 'Eight principles of integrated pest management', *Agronomy for Sustainable Development*, 35(4), pp. 1199–1215.

Bektas, Y. and Eulgem, T. (2015) 'Synthetic plant defense elicitors', *Frontiers in Plant Science*, 5, p. 804. doi: 10.3389/fpls.2014.00804

Ben Issa, R., Gautier, H. and Gomez, L. (2017) 'Influence of neighbouring companion plants on the performance of aphid populations on sweet pepper plants under greenhouse conditions', *Agricultural and Forest Entomology*, 19(2), pp. 181–191.

Birkett, M. A. *et al.* (2000) 'New roles for cis-jasmone as an insect semiochemical and in plant defense', *Proceedings of the National Academy of Sciences of the United States of America*, 97(16), pp. 9329–9334. doi: 10.1073/pnas.160241697

Bottrell, D.G. and Schoenly, K.G., 2018. Integrated pest management for resource-limited farmers: challenges for achieving ecological, social and economic sustainability. *The Journal of Agricultural Science*, *156*(3), pp. 408–426.

Boughton, A. J., Hoover, K. and Felton, G. W. (2006) 'Impact of chemical elicitor applications on greenhouse tomato plants and population growth of the green peach aphid, Myzus persicae', *Entomologia Experimentalis et Applicata*, 120(3), pp. 175–188.

Brillada, C. *et al.* (2013) 'Metabolic engineering of the C16 homoterpene TMTT in Lotus japonicus through overexpression of (E,E)-geranyllinalool synthase attracts generalist and specialist predators in different manners', *New Phytologist*, 200, pp. 1200–1211. doi: 10.1111/nph.12442

Bruce, T. J. A. *et al.* (2003) 'cis-jasmone treatment induces resistance in wheat plants against the grain aphid, Sitobion avenae (Fabricius) (Homoptera: Aphididae)', *Pest Management Science*, 59(9), pp. 1031–1036. doi: 10.1002/ps.730

Cao, H. H., Wang, S. H. and Liu, T. X. (2014) 'Jasmonate- and salicylate-induced defenses in wheat affect host preference and probing behavior but not performance of the grain aphid, Sitobion avenae', *Insect Science*, 21(1), pp. 47–55. doi: 10.1111/1744-7917.12023

Cipollini, D. and Heil, M. (2010) 'Costs and benefits of induced resistance to herbivores and pathogens in plants', *CAB Reviews: Perspectives in Agriculture, Veterinary Science, Nutrition and Natural Resources*, 5. doi: 10.1079/PAVSNNR20105005

Cook, S. M., Khan, Z. R. and Pickett, J. A. (2007) 'The use of push-pull strategies in integrated pest management', *Annual Review of Entomology*, 52, pp. 375–400.

Dardouri, T. *et al.* (2017) 'How French marigold (Tagetes patula L.) volatiles can affect the performance of green peach aphid', *IOBC-WPRS Bull*, 123, pp. 71–78.

Dardouri, T. *et al.* (2019) 'Repellence of Myzus persicae (Sulzer): evidence of two modes of action of volatiles from selected living aromatic plants', *Pest Management Science*, 75(6), pp. 1571–1584.

De Moraes, C. M., Mescher, M. C. and Tumlinson, J. H. (2001) 'Caterpillar-induced nocturnal plant volatiles repel conspecific females', *Nature*, 410(6828), pp. 577–580.

Deguine, J.-P. *et al.* (2021) 'Integrated pest management: good intentions, hard realities. A review', *Agronomy for Sustainable Development*, 41(3), pp. 1–35.

Denkyirah, E.K., Okoffo, E.D., Adu, D.T., Aziz, A.A., Ofori, A. and Denkyirah, E.K., 2016. Modeling Ghanaian cocoa farmers' decision to use pesticide and frequency of application: the case of Brong Ahafo Region. *SpringerPlus*, *5*, pp. 1–17.

Després, L., David, J. and Gallet, C. (2007) 'The evolutionary ecology of insect resistance to plant chemicals', *Trends in Ecology & Evolution*, 22(6), pp. 298–307. doi: 10.1016/j.tree.2007.02.010

Disi, J. O. *et al.* (2017) 'cis-Jasmone primes defense pathways in tomato via emission of volatile organic compounds and regulation of genes with consequences for Spodoptera exigua oviposition', *Arthropod-Plant Interactions*, 11(4), pp. 591–602.

Dutta, P., Bhattacharyya, A. and Kumari, A., 2023. 'Innovative integrated pest management paradigm for sustainable crop production with special reference to North East India', in *Integrated pest management in diverse cropping systems*. Apple Academic Press, pp. 61–90.

Eizi, Y. (2021) 'Augmentative biological control in greenhouses in Japan', *CABI Reviews*.

Ekesi, S., Dimbi, S. and Maniania, N. K. (2007) 'The role of entomopathogenic fungi in the integrated management of fruit flies (Diptera: Tephritidae) with emphasis on species occurring in Africa', in *Use of entomopathogenic fungi in biological pest management*, Editors: S. Ekesi, N. K. Maniania, Research Signpost, pp. 239–274.

Erbilgin, N. *et al.* (2006) 'Exogenous application of methyl jasmonate elicits defenses in Norway spruce (Picea abies) and reduces host colonization by the bark beetle Ips typographus', *Oecologia*, 148(3), pp. 426–436.

Foglar, H., Malausa, J. C. and Wajnberg, E. (1990) 'The functional response and preference ofmacrolophus caliginosus [heteroptera: Miridae] for two of its prey: Myzus persicae andtetranychus urticae', *Entomophaga*, 35(3), pp. 465–474.

Gao, Y., Zhang, X., Lu, J., Wu, L. and Yin, S., 2017. Adoption behavior of green control techniques by family farms in China: Evidence from 676 family farms in Huang-huai-hai Plain. *Crop Protection*, *99*, pp. 76–84.

Gepts, P. (2002) 'A comparison between crop domestication, classical plant breeding, and genetic engineering', *Crop Science*, 42(6), pp. 1780–1790.

Gozzo, F. and Faoro, F. (2013) 'Systemic acquired resistance (50 years after discovery): moving from the lab to the field', *Journal of Agricultural and Food Chemistry*, 61, pp. 12473–12491. doi: 10.1021/jf404156x

Haggblade, S., Diarra, A. and Traoré, A., 2022. Regulating agricultural intensification: Lessons from West Africa's rapidly growing pesticide markets. *Development Policy Review*, *40*(1), p. e12545.

Heil, M. (2008) 'Indirect defence via tritrophic interactions', *New Phytologist*, 178(1), pp. 41–61. doi: 10.1111/j.1469-8137.2007.02330.x

Holopainen, J. K. *et al.* (2009) 'Potential for the use of exogenous chemical elicitors in disease and insect pest management of conifer seedling roduction', *The Open Forest Science Journal*, 2(1), pp. 17–24.

Howe, G. A. and Jander, G. (2008) 'Plant immunity to insect herbivores', *Annual Review of Plant Biology*, 59, pp. 41–66. doi: 10.1146/annurev.arplant.59.032607.092825

Jaber, L. R. and Araj, S.-E. (2018) 'Interactions among endophytic fungal entomopathogens (Ascomycota: Hypocreales), the green peach aphid Myzus persicae Sulzer (Homoptera: Aphididae), and the aphid endoparasitoid Aphidius colemani Viereck (Hymenoptera: Braconidae)', *Biological Control*, 116, pp. 53–61.

Jacobsen, B. J., Zidack, N. K. and Larson, B. J. (2004) 'The role of Bacillus-based biological control agents in integrated pest management systems: plant diseases', *Phytopathology*, 94(11), pp. 1272–1275.

Kang, B. R. *et al.* (2018) 'Dual biocontrol potential of the entomopathogenic fungus, Isaria javanica, for both aphids and plant fungal pathogens', *Mycobiology*, 46(4), pp. 440–447.

Kappers, I. F. *et al.* (2005) 'Genetic engineering of terpenoid metabolism attracts bodyguards to Arabidopsis', *Science*, 309(5743), pp. 2070–2072. doi: 10.1126/science.1116232

Karban, R. and Myers, J. H. (1989) 'Induced plant responses to herbivory', *Annual Review of Ecology and Systematics*, 20(1), pp. 331–348. doi: 10.1146/annurev.es.20.110189.001555

Lee, W. W. *et al.* (2015) 'Screening and evaluation of entomopathogenic fungi against the green peach aphid, Myzus persicae, using multiple tools', *Journal of Asia-Pacific Entomology*, 18(3), pp. 607–615.

Lewis, W. J. *et al.* (1997) 'A total system approach to sustainable pest management', *Proceedings of the National Academy of Sciences*, 94(23), pp. 12243–12248.

Liu, J. *et al.* (2019) 'Preference of the aphid Myzus persicae (Hemiptera: Aphididae) for tobacco plants at specific stages of potato virus Y infection', *Archives of Virology*, 164(6), pp. 1567–1573.

Lu, J. *et al.* (2015) 'Induced jasmonate signaling leads to contrasting effects on root damage and herbivore performance', *Plant Physiology*, 167(March), pp. 1100–1116. doi: 10.1104/pp.114.252700

Ma, Z. *et al.* (2013) 'Chitosan and oligochitosan enhance the resistance of peach fruit to brown rot', *Carbohydrate Polymers*, 94(1), pp. 272–277.

Matthes, M. *et al.* (2011) 'Emerging roles in plant defense for cis-jasmone-induced cytochrome P450 CYP81D11', *Plant Signaling & Behavior*, 6(4), pp. 563–565.

Mishra, M. *et al.* (2015) 'Ecological turmoil in evolutionary dynamics of plant–insect interactions: defense to offence', *Planta*, 242(4), pp. 761–771.

Mithöfer, A. and Boland, W. (2012) 'Plant defense against herbivores: chemical aspects', *Annual Review of Plant Biology*, 63, pp. 431–450. doi: 10.1146/annurev-arplant-042110-103854

Naranjo, S. E. (2001) 'Conservation and evaluation of natural enemies in IPM systems for Bemisia tabaci', *Crop Protection*, 20(9), pp. 835–852.

Oluwafemi, S. *et al.* (2013) 'Priming of production in maize of volatile organic defence compounds by the natural plant activator cis-jasmone', *PloS One*, 8(6), p. e62299. doi: 10.1371/journal.pone.0062299

Park, H. W. *et al.* (2013) 'Infectivity of entomopathogenic nematode Steinernema carpocapsae Pocheon strain (Nematoda: Steinernematidae) on the green peach aphid Myzus persicae (Hemiptera: Aphididae) and its parasitoids', *Biocontrol Science and Technology*, 23(6), pp. 637–645.

Prakash, A. and Rao, J. (2018) *Botanical pesticides in agriculture*. CRC Press.

Ren, G. *et al.* (2015) 'Potato virus Y-infected tobacco affects the growth, reproduction, and feeding behavior of a vector aphid, Myzus persicae (Hemiptera: Aphididae)', *Applied Entomology and Zoology*, 50(2), pp. 239–243.

Rijal, J.P., Regmi, R., Ghimire, R., Puri, K.D., Gyawaly, S. and Poudel, S., 2018. Farmers' knowledge on pesticide safety and pest management practices: A case study of vegetable growers in Chitwan, Nepal. *Agriculture*, *8*(1), p. 16.

Rodriguez-Saona, C. *et al.* (2011) 'Field responses of predaceous arthropods to methyl salicylate: A meta-analysis and case study in cranberries', *Biological Control*, 59(2), pp. 294–303. doi: 10.1016/j.biocontrol.2011.06.017

Singh, H. and Joshi, N. (2020) 'Management of the aphid, Myzus persicae (Sulzer) and the whitefly, Bemisia tabaci (Gennadius), using biorational on capsicum under protected cultivation in India', *Egyptian Journal of Biological Pest Control*, 30(1), pp. 1–9.

Skinner, M., Parker, B. L. and Kim, J. S. (2014) 'Role of entomopathogenic fungi in integrated pest management', *Integrated Pest Management*, pp. 169–191.

Sobhy, I. S. *et al.* (2014) 'The prospect of applying chemical elicitors and plant strengtheners to enhance the biological control of crop pests', *Philosophical Transactions of the Royal Society B: Biological Sciences*, 369(1639), p. 20120283. doi: 10.1098/rstb.2012.0283

Sobhy, I. S. *et al.* (2017) 'cis-Jasmone elicits aphid-induced stress signalling in Potatoes', *Journal of Chemical Ecology*, 43(1), pp. 39–42. doi: 10.1007/s10886-016-0805-9

Stenberg, J. A. (2017) 'A conceptual framework for integrated pest management', *Trends in Plant Science*, 22(9), pp. 759–769.

Sunderland, K. and Samu, F. (2000) 'Effects of agricultural diversification on the abundance, distribution, and pest control potential of spiders: a review', *Entomologia Experimentalis et Applicata*, 95(1), pp. 1–13.

Symondson, W. O. C., Sunderland, K. D. and Greenstone, M. H. (2002) 'Can generalist predators be effective biocontrol agents', *Annual Review of Entomology*, 47(1), pp. 561–594.

Tiwari, S., Sharma, S. and Wratten, S. D. (2020) 'Flowering alyssum (Lobularia maritima) promote arthropod diversity and biological control of Myzus persicae', *Journal of Asia-Pacific Entomology*, 23(3), pp. 634–640.

Tomilova, O. G. *et al.* (2021) 'Early physiological response of potato plants to entomopathogenic fungi under hydroponic conditions', *Horticulturae*, 7(8), p. 217.

Wei, T. *et al.* (2021) 'Effect of exogenous silicon and methyl jasmonate on the alleviation of cadmium-induced phytotoxicity in tomato plants', *Environmental Science and Pollution Research*, 28(37), pp. 51854–51864.

Whipps, J. M. and Lumsden, R. D. (2001) 'Commercial use of fungi as plant disease biological control agents: status and prospects', in *Fungi as biocontrol agents: progress, problems and potential*, Editors: T. M. Butt, C. Jackson, N. Magan, pp. 9–22.

Xiao, L. *et al.* (2019) 'Herbivore-specific induction of indirect and direct defensive responses in leaves and roots', *AoB Plants*, 11(1), p. plz003.

6 Wild Ancestors of the Potato as Potential Sources of Resistance against Aphid *Myzus persicae*

6.1 INTRODUCTION

Genetic resources for resistance against pests are limited within the current cultivars of potato, resulting in the susceptibility of these cultivars to pests and disease. In contrast, the wild relatives of the potato grow in a wide range of environments that make them genetically more diverse than the domesticated cultivars of *Solanum* species. Furthermore, they have not been through genetic bottlenecks that restrict genetic variation when plant material is selected for breeding. Commercially available potato cultivars are susceptible to a wide range of insect pests. Factors responsible for plant growth and development can affect the level of resistance exhibited by the plant species (Panda and Khush, 1995; Smith, 2005; Gaillard *et al.*, 2018). Plants growing in fluctuating environmental conditions face difficulties attaining the full genetic potential required for growth and defense (Iqbal *et al.*, 2021). For example, the presence or absence of light not only affects the growth of plants but also has an impact on plant defense by affecting the biochemistry of plants and defense against attackers.

The deployment of a defense strategy is critical for plant survival but plant defense activation comes at the expense of plant growth (Huot *et al.*, 2014). In nature, plants encounter a large number of pests; in order to grow and combat the pests, the plant has evolved a sophisticated mechanism to adjust its resources to both grow and defend the pests. However, most crop-breeding programs focus on growth-related traits that lead to a loss of genetic diversity and make them vulnerable to pests by compromising their defense capacity. Factors that make commercially available crops susceptible to insect pests and diseases include a lack of stability in a changing environment, down-regulation of defense-related genes and where there has been intense use of insecticides crop resistance traits are not needed and therefore they may have been lost (Peterson *et al.*, 1992; Chaudhary *et al.*, 2008). Wild relatives of crop plants develop in a comprehensive environment that makes them genetically more diverse (Bradshaw, Bryan and Ramsay, 2006; HeřmanoVá, Bárta and Čurn, 2007). These wild cultivars could be a potential source of resistance genes and can be utilized in plant breeding programs (Leppik, 1970; Lenne and Wood, 1991;

DOI: 10.1201/9781003400974-6

Hajjar and Hodgkin, 2007). Genetic and environmental factors regulate the content of secondary metabolites in different wild and cultivated species of plants. Secondary metabolites play a significant role in plant resistance and genes encoding that can be transferred to progenies during crop breeding (Wink, 1988; Ode, 2006; Züst and Agrawal, 2016). Phytochemicals can be used in strategies to lower the attractiveness such as changes in the taste of the plant (making it unpalatable for herbivores) that consequently reduce the level of herbivore attack (El-Sayed *et al.*, 2006; Koul, 2008; Ratnadass *et al.*, 2012; Gregg, Del Socorro and Landolt, 2018).

In this chapter, we aimed to identify the resistance level of wild generations of *Solanum stoloniferum* potatoes, compared with cultivated ones, against the aphid *M. persicae*. In addition, we also recorded the behavioral responses of parasitoid *Diaeretiella rapae* a common natural enemy of the peach-potato aphid to odours collected from the potato plants. The resistance of wild relatives of several crop plant species is well documented (Prescott-Allen and Prescott-Allen, 1986; Cooper, Spillane and Hodgkin, 2001; Hajjar and Hodgkin, 2007; Powles and Yu, 2010; Moore, 2015); however, previous studies have not investigated the lines that we tested here.

6.2 AIM AND OBJECTIVES

- To test the performance of *M. persicae* on different lines of potato species.
- To identify the difference between the blend of volatile compounds released from different lines of potato species.
- To test the behavioral response of *M. persicae* and *D. rapae* to odor collected from different lines of potato species.

6.3 APHID PERFORMANCE BIOASSAY

The performance of *M. persicae* was assessed on the wild potato accessions and compared with cultivated potato species. There were two separate series of experiments with different plants; the first series recorded observation after 48 hours and the second recorded observations after 96 hours. Fresh plants and aphids were used in each replicate observation in each experiment. In each replicate 10 adult alate *M. persicae* were placed in a clip cage (2.5 cm diameter, Bioquip Product Inc. USA), which was attached to the lower surface of plant leaves (Sobhy *et al.*, 2020) (Figure 6.1). Two clip cages were placed on each plant. Ten replicates, on separate plants, were performed for each genotype. To assess the survival and fecundity of aphids, plants were left undisturbed in a controlled environment room (25 °C ± 2 °C, 37 % RH, 16 h L: 8 h D photoperiod). Plants were assessed after the 48-hour (series 1) or 96-hour (series 2) period. For assessment, leaves containing the cages were cut and cages were removed without losing any aphids. The number of live adults and nymphs produced were recorded.

After 48 hours, there was a significant reduction in adult *M. persicae* survival on wild potato lines in clip cage experiments. The number of live adult *M. persicae* on Desiree (mean = 9.5) was up to more than three-fold higher compared to the number on wild plants (18333 mean = 2.6) ($F_{3,76} = 63.732$; $P < 0.001$; Figure 6.2a). A similar pattern was observed in clip cage experiments after 96 hours. All wild lines had high

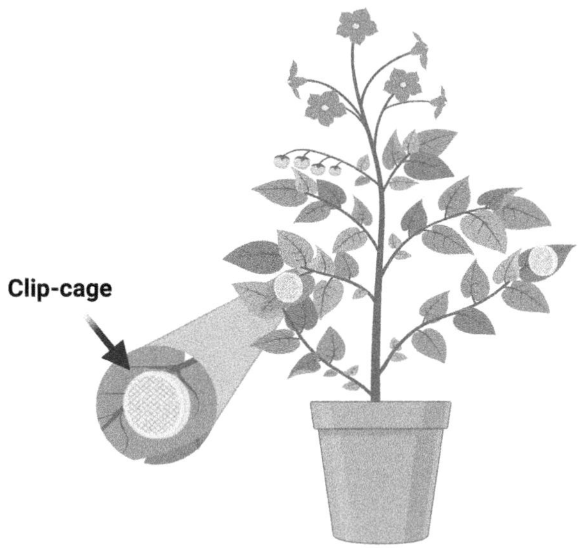

FIGURE 6.1 A set up showing clip-cages attached to a potato plant to assess the performance of aphid *Myzus persicae*.

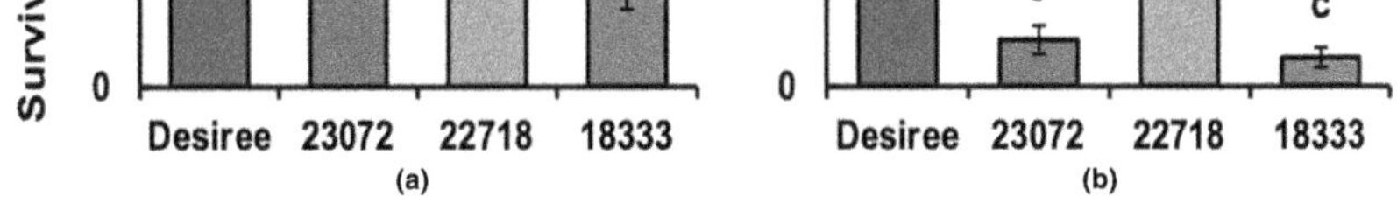

FIGURE 6.2 Performance of *Myzus persicae* on cultivated (*Solanum tuberosum*. cv. Desiree) and wild (*Solanum stoloniferum*) potato lines. Mean number (± standard error) of surviving adult aphids *M. persicae* after 48 hours (a) and 96 hours (b). Different letters indicate statistically significant differences among plant species (*F*-test; $P < 0.05$), based on the Holm–Sidak method (one-way ANOVA).

aphid mortality with a less than 35% rate of survival. Accessions 18333 and 23072 had 8% and 13% survival of aphids, respectively ($F_{3,76} = 46.299$; $P < 0.001$; Figure 6.2b). In contrast, Desiree had the least mortality (mean = 7.60) with more than 75% survival rate after 96 hours.

There was a significant reduction in nymph production on wild lines of *Solanum stoloniferum* across both time points compared to cultivated potatoes. Mean larviposition on wild accessions was significantly lower after 48 hours, decreasing by 85% from

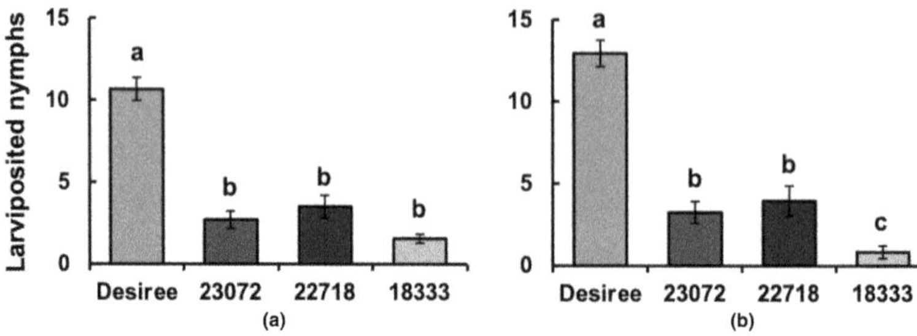

FIGURE 6.3 Performance of *Myzus persicae* on cultivated (*Solanum tuberosum*. cv. Desiree) and wild (*Solanum stoloniferum*) potato lines. Mean number (± SE) of larviposited nymphs of *M. persicae* after 48 h (a) and 96 h (b). Different letters indicate statistically significant differences among plant species (*F*-test; $P < 0.05$), based on the Holm–Sidak method (one-way ANOVA).

10.65 on Desiree plants to 1.55 on wild plants ($F_{3,76} = 48.428$; $P < 0.001$) Figure 6.3a. In wild accessions, larviposition was reduced most on 18333 and least on 22718. Similarly, mean larviposition on all wild accessions was significantly reduced after 96 hours, decreasing by 93% from 12.95 on Desiree to 0.85 on 18333 ($F_{3,76} = 50.739$; $P < 0.001$) Figure 6.3b.

6.4 AIR ENTRAINMENT

Headspace sampling allows samples to be collected with the chemicals released by the plant in ratios similar to those found in nature. In this project, an air entrainment was carried out following a procedure adapted from Agelopoulos *et al*., (1999). During volatile collection, the plant was kept inside a bag (35 × 43 cm; Bacofoil, UK) to collect the plant volatiles. The bag was partially sealed, so that only the volatiles produced by the plants were collected. Prior to entrainments, bags were baked in an oven (Heraeus, Thermo Electron corporation, Mark Biosciences, UK) at 120°C overnight. The Porapak Q filters (0.05 g, 60/80 mesh; Supelco, Bellefonte, PA, USA) were rinsed with diethyl ether and conditioned before use. Plants with five true leaves were enclosed in bags individually. Each bag was open at the bottom and closed at the top. An outlet hole was made in the upper part of the bag to connect the Porapak Q filter, whereas the bag was closed by attaching a rubber band around the pot (Figure 6.4). Charcoal filtered air was pumped in at 600 mL min^{-1}, and sampled air was pulled out at 400 mL min^{-1} through a Porapak Q filter in which the plant volatiles were trapped. To avoid the entry of unfiltered air, positive pressure was maintained by using differing in flow rates. Connections were made with 1.6 mm (i.d.) polytetrafluoroethylene (PTFE) tubing (Alltech Associates Inc., Lancashire, UK) with Swagelok brass ferrules and fitting (North London Valve Co., London, UK) and sealed with PTFE tape (Gibbs & Dandy Ltd., Luton, UK) (Figure 6.3). Volatile collection was done for a period of 48 hours, after which the Porapak filters were eluted with 0.5 mL of diethyl ether, into sample vials (Supelco, 2 mL, PTFE/silicone) and stored at −20°C in a freezer (Lec Medical, UK) for use in olfactometer bioassays and chemical analysis.

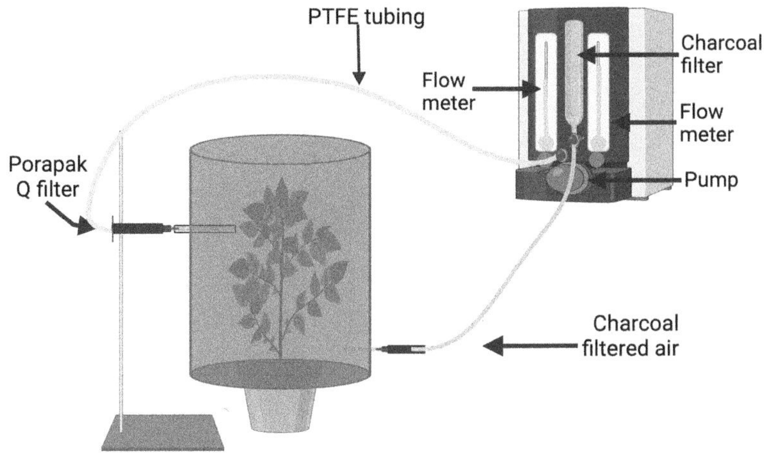

FIGURE 6.4 Illustration showing a set up used for plant volatile collection.

6.5 BEHAVIORAL BIOASSAY

The behavioral response of alate *M. persicae* and *D. rapae* to potato volatile organic compounds (VOCs) was investigated using a Perspex 4-arm olfactometer in a controlled environment room (24°C ± 2, 30 ± 2% RH). At the top of the olfactometer, the central area contained a hole into which a single alate *M. persicae* or female *D. rapae* were introduced, which was connected to a low-pressure air pump. Air was pulled out at 200 mL/min from the center of the olfactometer by a vacuum pump with a layer of muslin to prevent access by an aphid during the bioassays. All replicates were done under uniform illumination. The olfactometer arena was split into five areas; four areas by each arm (one or two treatment(s) and three, or two, control arms) and a central area (Webster *et al.*, 2010) (Figure 6.5). Each replicate was run for 12 min., and after every 3 min., the position of the olfactometer was rotated clockwise by 90° so as to eliminate bias. Time spent by the insect in each arm was recorded using a software program OLFA (F. Nazzi Udine, Italy). Ten replicates were done for each insect. Filter paper (Whatman Filter Paper, Buckinghamshire, UK) strips (cut to 5 × 20 mm) were treated with an aliquot (10 μL) of the test solution, applied using a micropipette (Drummond 'microcaps'; Drummond Scientific Co., USA). One arm was assigned to the collected VOCs from the potato plants, whereas three control vessels were treated similarly with the same volume of solvent (diethyl ether) on the filter paper strips (wild potato vs solvent and Desirée vs solvent). If an insect remained motionless for the first 2 min. of a replicate, that replicate was discarded. All bioassays were performed between 10:00 and 13:00.

6.5.1 Aphid Behavioral Response

In an olfactometer bioassay, *M. persicae* were repelled by the volatiles of wild accessions, whereas they spent more time in the treated zone for Desiree ($P = 0.19$). Wild accessions had a significant repellent effect on *M. persicae*. Accessions 18333 and 23072 showed a significant repellent effect on *M. persicae* with P values of 0.013 and 0.018, respectively (Figure 6.6).

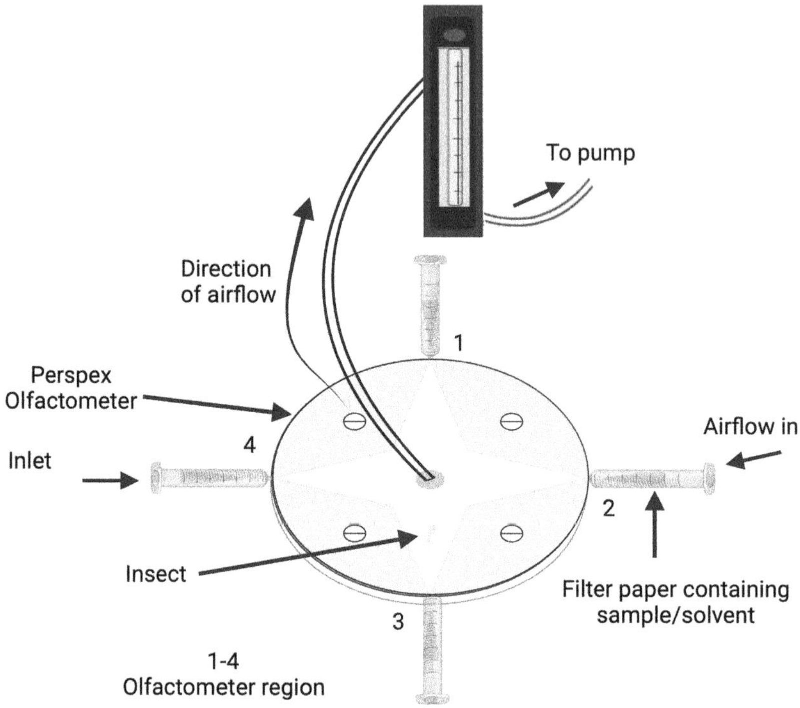

FIGURE 6.5 A generalized illustration showing four-arm olfactometer with odour inlets.

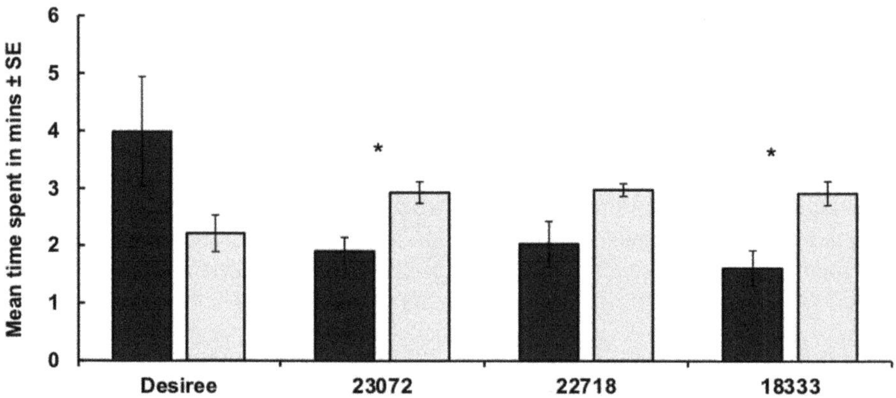

FIGURE 6.6 Behavioral responses of *Myzus persicae* to headspace volatiles from cultivated (*Solanum tuberosum*. cv. Desiree) and wild (*Solanum stoloniferum*) potato lines in a four-arm olfactometer bioassay. Each insect was given 12 min. to make a choice between one arm of plant volatiles (black bars) versus three solvent diethyl ether (DEE) arms (grey bars). The values shown are mean time spent in the arm ± SE ($n = 10$). Asterisks ($*0.01 \leq P \leq 0.05$) above bars indicate statistically significant differences based on a paired t-test (one tail).

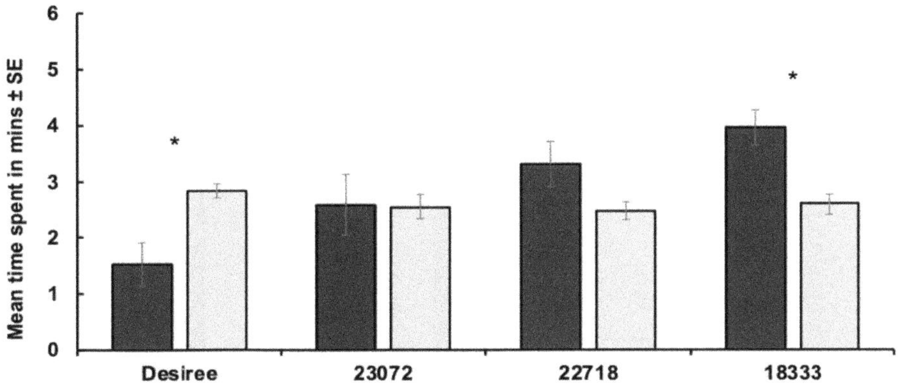

FIGURE 6.7 Behavioral responses of *Diaeretiella rapae* to headspace volatiles from cultivated (*Solanum tuberosum*. cv. Desiree) and wild (*Solanum stoloniferum*) potato lines in a four-arm olfactometer bioassay. Each insect was given 12 min. to make a choice between one arm of plant volatiles (black bars) versus three solvent diethyl ether (DEE) arms (grey bars). The values shown are mean time spent in the arm ± SE ($n = 10$). Asterisks (*$0.01 \leq P \leq 0.05$) above bars indicate statistically significant differences based on a paired t-test (one tail).

6.5.2 Parasitoid Behavioral Response

In contrast to *M. persicae*, odour collected from wild accessions attracted parasitoids and odor collected from Desiree repelled the parasitoid. *D. rapae* showed a preference for the blend of volatiles emitted from wild accessions and spent more time in the zone treated with volatiles collected from wild accessions. Accession 18333 was the only wild accession that had a significant attractant effect on *D. rapae* ($P = 0.012$). *D. rapae* spent significantly less time ($P = 0.016$) in the olfactometer zone treated with volatiles collected from Desiree (Figure 6.7).

6.6 VOLATILE ANALYSIS

GC-MS analysis of entrainment samples of wild (18333, 22718, 23072) and cultivated accessions of potato revealed 23 detectable organic compounds belonging to different functional classes (alcohol, aldehydes, benzenoids, ketones, terpenes). There were significant quantitative differences (3 to 7 fold) in the total emitted volatiles compared to Desiree plants ($F_{3, 12} = 61.20$; $P < 0.001$) (Figure 6.8). In addition, most volatile compounds in the above-mentioned VOC groups were emitted from wild accessions in significantly higher amounts compared to Desiree plants. The difference in volatile emissions from wild and Desiree accessions may impact the behavioral responses of *M. persicae* and its natural enemy *D. rapae*.

Table 6.1 shows the details of the compounds identified from wild potato and cultivated accessions of potato. Among wild accessions, 23072 and 18333 accessions released high amount of volatile compounds while 22718 showed comparatively low volatile emissions. Volatile compounds p-Cymon-7-ol, 4-ethyl-benzaldehyde, MeSA,

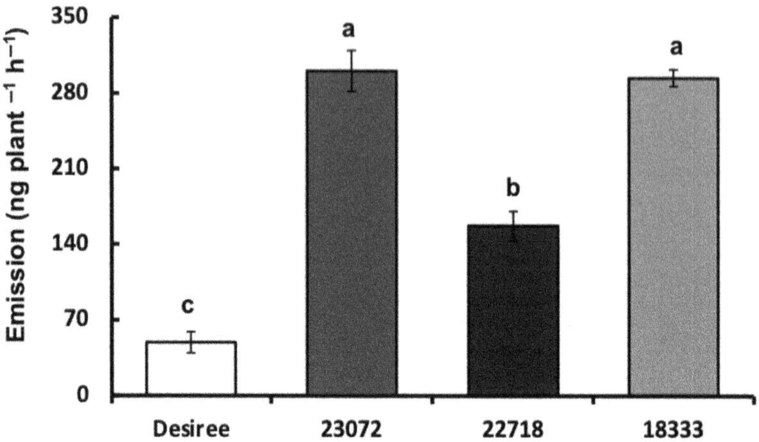

FIGURE 6.8 Total amounts (mean nanograms plant^{-1} h^{-1} ± SE) of identified volatile organic compounds (VOCs) emitted from cultivated (*Solanum tuberosum*, cv. *Desiree*) and wild (*Solanum stoloniferum*) potato lines. Different letters indicate statistically significant differences among plant species (*F*-test; $P < 0.05$), based on the Holm-Sidak method (one-way ANOVA).

D-Limonene, Benzothiazole, MHO, α-Copaene, β-elemene, trans-α-Bergamotene, (*E*)-β-Farnesene, and Germacrene D were the main compounds released by wild potato lines, while (*E*)-4,8-dimethyl-1,3,7-nonatriene (DMNT) was the main compound released by Desiree in significantly high amounts. In particular, the main compounds released by 23072 were p-Cymon-7-ol, 4-ethyl-benzaldehyde, D-Limonene, and Germacrene D. Accession 18333 released 11 main compounds that include p-Cymon-7-ol, 4-ethyl-benzaldehyde, Benzothiazole, D-Limonene, TMTT, α-Copaene, β-elemene, trans-α-Bergamotene and (*E*)-β-Farnesene. While a marked increase in emission of about 4 volatiles, including MeSA, β-Myrcene, D-Limonene and Caryophyllene, was recorded in 22718.

Under each chemical class, VOCs are ordered in accordance with their increasing retention time in a gas chromatograph and Kovats index. # [DMNT: (*E*)-4,8-dimethyl-1,3,7-nonatriene; MeSA: methyl salicylate; MHO: 6-methyl-5-hepten-2-one: ND: Not Detected]. VOCs were tentatively identified based on spectra, Kovats retention index and NIST 17 library matches. KI: Kovats index determined on the intermediately nonpolar HP5-MS column (https://webbook.nist.gov/; http://www.pherobase.com/). Different letters in the same row indicate significant differences between potato lines (One way ANOVA; $P < 0.05$). *P*-values in bold indicate significant difference.

PCA of the VOCs showed that the first two principal components (PCs) accounted for 54.3% of the total variation in the volatile data (Figure 6.9). Hence, these two PCs illustrated most of the variation in the data of likely biological relevance. A clear separation based on the second principal component (PC2) is visible between the volatile profiles of wild (18333, 22718, 23072) in one cluster and cultivated (Desiree)

TABLE 6.1
Emission (in ng per plant -1 h-1; mean ± SE; n = 3) of Volatiles Released by Cultivated (*Solanum tuberosum*. cv. Desiree) and Wild (*Solanum stoloniferum*) Potatoes Lines

		Solanum stoloniferum			*S. tuberosum*	
Plant Volatiles	**KI**	**23072**	**22718**	**18333**	**Desiree**	***P***
Alcohols						
Phenylethyl alcohol	1116	3.89±1.26	4.25±0.87	2.63±0.58	2.59±0.82	0.688
p-Cymen-7-ol	1289	119.27±11.45[a]	17.52±6.48[c]	64.32±19.07[b]	1.37±0.38[d]	**<0.001**
Aldehydes						
4-ethyl-benzaldehyde	1122	63.49±8.63[a]	7.31±3.19[b]	21.60±6.47[b]	1.11±0.10[b]	**<0.001**
Benzenoids						
MeSA #	1192	7.81±0.93[b]	26.08±6.08[a]	4.39±0.51[b]	3.25±0.37[b]	**0.002**
Benzothiazole	1229	1.05±0.52[b]	7.59±1.42[a]	10.07±0.93[a]	1.00±0.09[b]	**0.002**
Ketones						
MHO	989	13.15±0.65	4.39±1.66	10.69±2.79	0.49±0.24	**0.024**
Monoterpenes						
β-Myrcene	992	3.65±0.41	5.95±1.38	3.75±0.42	1.84±0.61	0.062
p-Cymene	1026	1.79±0.14	1.86±0.20	1.78±0.17	2.56±0.19	0.104
D-Limonene	1030	20.39±2.09[a]	23.99±2.86[a]	19.83±1.98[a]	0.71±0.25[b]	**0.003**
Linalool	1099	8.71±2.76[a]	9.37±1.58[a]	8.83±0.60[a]	2.52±0.76[b]	0.083
Homoterpenes						
DMNT#	1116	0.76±0.08[b]	2.15±0.18[b]	1.15±0.22[b]	11.26±3.24[a]	**0.045**
TMTT#	1577	2.62±0.38[b]	4.10±0.84[b]	23.68±3.89[a]	0.79±0.08[b]	**<0.001**
Sesquiterpenes						
β-Cubebene	1351	10.27±2.46	5.62±2.17	4.09±0.86	7.38±3.86	0.497
α-Copaene	1376	3.70±0.58[c]	7.34±0.64[b]	15.12±1.08[a]	1.89±0.59[c]	**<0.001**
β-elemene	1391	4.07±1.67[b]	2.67±0.18[b]	40.52±14.16[a]	1.02±0.04[b]	**0.014**
Longifolene	1402	3.29±0.72	2.72±0.76	2.72±0.69	ND	0.858
Caryophyllene	1419	5.49±0.67	6.89±0.59	4.72±0.28	7.92±2.47	0.485
trans-α-Bergamotene	1435	1.35±0.14[b]	3.54±0.32[b]	20.69±7.69[a]	0.81±0.08[b]	**0.024**
(*E*)-β-Farnesene	1457	1.61±0.11[b]	4.34±0.38[ab]	10.13±3.68[a]	1.31±0.13[b]	**0.045**
Germacrene D	1481	21.23±5.45[a]	1.49±0.08[c]	6.68±1.70[b]	1.25±0.21[c]	**0.005**
β-Selinene	1486	1.41±0.09[b]	6.31±1.21[a]	7.27±0.63[a]	0.97±0.16[b]	**<0.001**
β-Bisabolene	1509	1.59±0.20[c]	2.78±0.26[b]	5.07±0.45[a]	0.83±0.05[c]	**<0.001**
Nerolidol	1566	1.68±0.05	2.11±0.07	6.49±2.71	0.87±0.15	0.101
Total emitted volatiles		**300.68±18.48**[a]	**157.19±13.89**[b]	**294.32±7.91**[a]	**49.42±9.76**[c]	**<0.001**

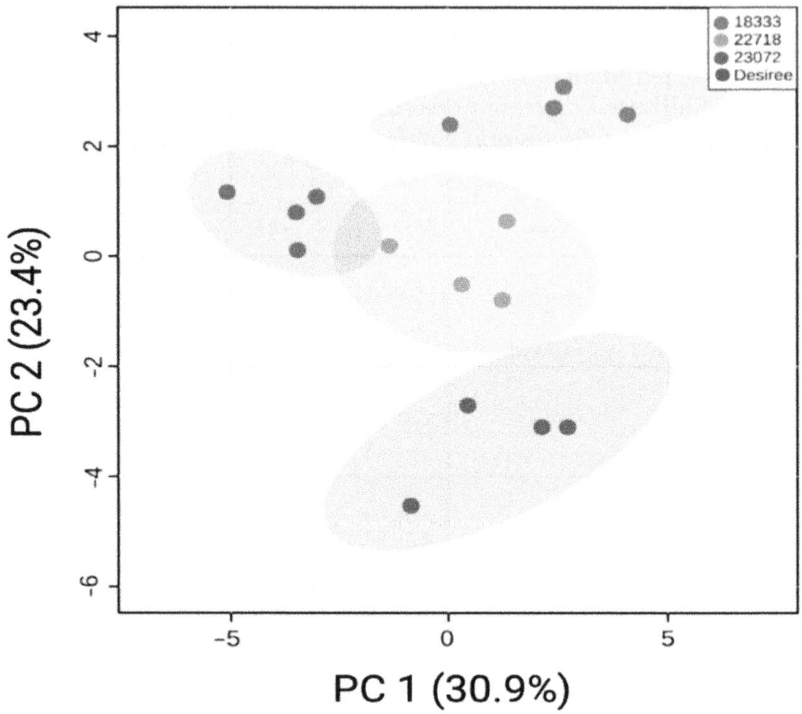

FIGURE 6.9 PCA of volatile compounds emitted by cultivated (*Solanum tuberosum.* cv. Desiree) and wild (*Solanum stoloniferum*) potato lines ($n = 4$) sampled for 48 hours. The score plot visualizes the ordination of collected samples according to the first two PCs based on the quantity of different volatiles emitted from different potato lines, with the percentage of the variation explained in parentheses. The ellipses show 95% confidence regions.

potato plants, whereas another separation but based on the first principal component (PC1) is obvious for the volatile profiles of 23072 and a cluster of 18333, 22718 and Desiree plants. The greatest loadings of PC2, in descending order, were for D-limonene (0.285), (E,E)-4,8,12-trimethyl-1,3,7,11-tridecatetraene (TMTT) (0.272), and p-cymen-7-ol (0.255), whereas the major loadings of PC1 were for ß-bisabolene (0.293), (E)-ß-farnesene (0.288), and trans-ʋ-bergamotene (0.285). This suggests that these VOCs, shown to contribute to PC1 and PC2, may impact the behaviour response of both *M. persicae* and *D. rapae*.

The compounds identified from wild potato odours (volatile blends) have been tested against various species of herbivore pests in previous studies. These studies suggested that the induced volatiles compounds have potential to negatively affect the plant pests, for instance; p-Cymon-7-ol found behaviorally active compounds when tested against tephritid pests and spring aphid *Kaburagia rhusicola* (Aluja *et al.*, 2020; Zhu *et al.*, 2020). The star anise *Illicium verum* extract with 4-ethyl-benzaldehyde as a component showed a repellent effect against maize weevil *Sitophilus zeamais* and *M. persicae* (Wei *et al.*, 2014; Shi-Guang *et al.*, 2017). Earlier studies reported that MeSA as an repellent compound that affects the behavior and oviposition of herbivorous pests (Ulland *et al.*, 2008; Snoeren *et al.*, 2010). Additionally, MeSA treatment of plant activates defensive signaling pathways (Riahi *et al.*, 2022). D-Limonene was effective against whiteflies, mealybugs and scale insects (Hollingsworth, 2005; Conboy *et al.*, 2019), β-elemene as a constituent of essential oil (*Tetradium glabrifolium*, Evodia rutaecarpa and *Zanthoxylum rhoifolium*) showed strong repellency and larvicidal activity against Asian tiger mosquito (*Aedes albopticus*), *Triboilum castaneum, Lesioderma serricorne Liposcelis bostrychophila* and *Bemisia tabaci* (Christofoli *et al.*, 2015; Liu *et al.*, 2015; Cao *et al.*, 2018), (*E*)-β-Farnesene induces changes in the behavior of the codling moth (Sutherland, Hutchins and Wearing, 1974; Yan *et al.*, 2003), In olfactometer bioassay, Germacrene D as a component of *Hemizigiya petiolate* affects *S. avenae* and *M. persicae* behaviour negatively (Bruce *et al.*, 2005).

Furthermore, the compounds released from wild potato accessions were also effective when tested with biocontrol agents i.e., predators and parasitoids. For example, 4-ethyl-benzaldehyde showed a positive effect on several natural enemies, including *Stethorus punctum*, *Cotesia plutellae* and *Trichogramma dendrolomi* (James, 2005; Yang *et al.*, 2016; Zhao *et al.*, 2022), Germacrene D was found attractive to predatory mite *Amblysehis cucumer* in olfactometer bioassay (Manjunatha *et al.*, 1998).

6.7 DISCUSSION

The peach-potato aphid *M. persicae* is a globally important pest for many economically important crops (van Emden *et al.*, 1969; Alyokhin *et al.*, 2013; van Emden and Harrington, 2017). Synthetic chemical pesticides used to be the main control option for this species, but increasing levels of pest resistance to these chemicals and safety concerns about food and the environment mean that new solutions are needed (Alyokhin *et al.*, 2013; Bass *et al.*, 2014; Singh *et al.*, 2021). Current research is focused on finding alternative methods to replace the use of traditional insecticides. Previous studies reported that plant secondary metabolites could provide a way to enhance plant resistance (Bennett and Wallsgrove, 1994; Schoonhoven *et al.*, 2005; Guerriero *et al.*, 2018). Plant resistance could be an important aspect in integrated pest management. The presence of secondary metabolites in a plant affects the survival and reproduction rate of insects (Schoonhoven *et al.*, 2005). Plants producing a high level of secondary metabolites have better defense against insects and pathogens and can be used to establish a sustainable agricultural system. Higher levels of disease and pest resistance are highly recommended in potato cultivars, but, of course, further to these properties, they must also retain the marketable yield and quality required for a modern cultivar to be successful (Bradshaw and Mackay, 1994).

The current study showed good levels of resistance in the wild potato lines tested. There was a significant difference in the susceptibility of wild and cultivated species of *Solanum*. The most resistant accession of wild potato in these experiments was 18333, followed by 23072 and 22718. The cultivated potato *S. tuberosum* cv. Desiree, used as a standard for comparison, had substantial nymph production after 96 hours, meaning it is much more susceptible. The wild accessions were shown to be resistant.

Among these two *Solanum* species, *S. tuberosum* (Desiree) was less resistant. Nymph production was reduced in wild accessions because there was a high mortality rate among adults and no more adults to produce the nymphs. In accessions 18333 and 23072, significant numbers of adults were dead after 96 hours.

The parasitoid *D. rapae* spent more time in the olfactometer arm treated with volatile samples collected from wild plants. In contrast, parasitoids spent significantly less time on Desiree plants. The olfactometer studies thus suggest that wild accessions help in recruiting natural enemies of aphids; however, field experiments are needed to fully confirm this. Recruitment of natural enemies is an important approach exhibited by the plants in defending the herbivores' attack (Price, 1987; Vet and Dicke, 1992; Gols, 2014). Plants that produce a large plethora of appropriate volatile compounds can successfully recruit a wide range of natural enemies (Turlings and Wäckers, 2004). However, the quantity and quality of released plant volatiles are critical. Sometimes compounds present in small quantities are more biologically active despite small quantities (Vet and Dicke, 1992). Quality, quantity, and the ratio of volatile compounds all play a crucial role in plant–insect interactions (Bruce *et al.*, 2010). Commercially available crop plants have been selected primarily to obtain an increased yield. This genetic manipulation through selection compromises their defense capacity by altering the interaction between plants, herbivores, and natural enemies (Chen, Gols and Benrey, 2015; Yolanda, Gols and Benrey, 2015).

The current study shows that there are promising sources of direct aphid resistance in wild potato germplasm. Furthermore, there is also evidence they could provide indirect resistance through the recruitment of natural enemies. The genetic makeup of these accessions creates chemical diversity in plants that affect the development of the aphid (Manrique-Carpintero *et al.*, 2013). In both of the *Solanum* species which were used in experiments, morphological differences could be seen; wild accessions had smaller leaves compared to Desiree. Although there could be some relation between aphids and leaf size, it is unlikely to explain the high mortality observed in the current study which is more likely due to the presence of toxic phytochemicals. The current research findings open up the prospect of breeding for aphid resistance by crossing cultivated and wild potatoes. However, we need to know what the chemicals conferring resistance are, and if they are safe for human consumption. It is also possible to have higher concentrations of protective chemicals in the leaves, but not in the tubers which are eaten.

REFERENCES

Agelopoulos, N., Birkett, M.A., Hick, A.J., Hooper, A.M., Pickett, J.A., Pow, E.M., Smart, L.E., Smiley, D.W., Wadhams, L.J. and Woodcock, C.M., 1999. Exploiting semiochemicals in insect control. *Pesticide Science*, *55*(3), pp. 225–235.

Aluja, M. *et al.* (2020) 'Host plant and antibiotic effects on scent bouquet composition of Anastrepha ludens and Anastrepha obliqua calling males, two polyphagous tephritid pests', *Insects*, 11(5), p. 309.

Alyokhin, A. *et al.* (2013) 'Evolutionary considerations in potato pest management', *Insect Pests of Potato Global Perspectives on Biology and Management*, 19, p. 543.

Bass, C. *et al.* (2014) 'The evolution of insecticide resistance in the peach potato aphid, Myzus persicae,' *Insect Biochemistry and Molecular Biology*, 51, pp. 41–51.

Bennett, R. N. and Wallsgrove, R. M. (1994) 'Secondary metabolites in plant defence mechanisms', *New Phytologist*, 127(4), pp. 617–633.

Bradshaw, J. E., Bryan, G. J. and Ramsay, G. (2006) 'Genetic resources (including wild and cultivated Solanum species) and progress in their utilisation in potato breeding', *Potato Research*, 49(1), pp. 49–65.

Bradshaw, J.E. and Mackay, G.R., 1994. Breeding strategies for clonally propagated potatoes. *Potato Genetics*, pp. 467–497.

Bruce, T. J. A. *et al.* (2005) 'Response of economically important aphids to components of Hemizygia petiolata essential oil', *Pest Management Science*, 61(11), pp. 1115–1121. doi: 10.1002/ps.1102

Bruce, T. J. A. *et al.* (2010) 'Is quality more important than quantity? Insect behavioural responses to changes in a volatile blend after stemborer oviposition on an African grass', *Biology Letters*, 6(3), pp. 314–317.

Cao, J.-Q. *et al.* (2018) 'Contact toxicity and repellency of the essential oils of Evodia lenticellata Huang and Evodia rutaecarpa (Juss.) Benth. leaves against three stored product insects', *Journal of Oleo Science*, 67(8), pp. 1027–1034.

Chaudhary, B. *et al.* (2008) 'Global analysis of gene expression in cotton fibers from wild and domesticated Gossypium barbadense', *Evolution & Development*, 10(5), pp. 567–582.

Chen, Y. H., Gols, R. and Benrey, B. (2015) 'Crop domestication and its impact on naturally selected trophic interactions', *Annual Review of Entomology*, 60, pp. 35–58.

Christofoli, M. *et al.* (2015) 'Insecticidal effect of nanoencapsulated essential oils from Zanthoxylum rhoifolium (Rutaceae) in Bemisia tabaci populations', *Industrial Crops and Products*, 70, pp. 301–308.

Conboy, N. J. A. *et al.* (2019) 'Companion planting with French marigolds protects tomato plants from glasshouse whiteflies through the emission of airborne limonene', *PloS One*, 14(3), p. e0213071.

Cooper, H. D., Spillane, C. and Hodgkin, T. (2001) *Broadening the genetic base of crop production*. CABI.

El-Sayed, A. M. *et al.* (2006) 'Potential of mass trapping for long-term pest management and eradication of invasive species', *Journal of Economic Entomology*, 99(5), pp. 1550–1564.

Gaillard, M. D. P. *et al.* (2018) 'Fine-tuning the 'plant domestication-reduced defense' hypothesis: specialist vs generalist herbivores', *New Phytologist*, 217(1), pp. 355–366.

Gols, R. (2014) 'Direct and indirect chemical defences against insects in a multitrophic framework', *Plant, Cell & Environment*, 37(8), pp. 1741–1752.

Gregg, P. C., Del Socorro, A. P. and Landolt, P. J. (2018) 'Advances in attract-and-kill for agricultural pests: beyond pheromones', *Annual Review of Entomology*, 63, pp. 453–470.

Guerriero, G. *et al.* (2018) 'Production of plant secondary metabolites: examples, tips and suggestions for biotechnologists', *Genes*, 9(6), p. 309.

Hajjar, R. and Hodgkin, T. (2007) 'The use of wild relatives in crop improvement: a survey of developments over the last 20 years', *Euphytica*, 156(1), pp. 1–13.

HeřmanoVá, V., Bárta, J. and Čurn, V. (2007) 'Wild potato species: characterization and biological potential for potato breeding', *Czech Journal of Genetics and Plant Breeding*, 43(3), pp. 73–81.

Hollingsworth, R. G. (2005) 'Limonene, a citrus extract, for control of mealybugs and scale insects', *Journal of Economic Entomology*, 98(3), pp. 772–779.

Huot, B. *et al.* (2014) 'Growth–defense tradeoffs in plants: a balancing act to optimize fitness', *Molecular Plant*, 7(8), pp. 1267–1287.

Iqbal, Z. *et al.* (2021) 'Plant defense responses to biotic stress and its interplay with fluctuating dark/light conditions', *Frontiers in Plant Science*, 12, p. 297.

James, D. G. (2005) 'Further field evaluation of synthetic herbivore-induced plan volatiles as attractants for beneficial insects', *Journal of Chemical Ecology*, 31(3), pp. 481–495.

Koul, O. (2008) 'Phytochemicals and insect control: an antifeedant approach', *Critical Reviews in Plant Sciences*, 27(1), pp. 1–24.
Lenne, J. M. and Wood, D. (1991) 'Plant diseases and the use of wild germplasm', *Annual Review of Phytopathology*, 29(1), pp. 35–63.
Leppik, E. E. (1970) 'Gene centers of plants as sources of disease resistance', *Annual Review of Phytopathology*, 8(1), pp. 323–344.
Liu, X. C. *et al.* (2015) 'Larvicidal activity of the essential oil from Tetradium glabrifolium fruits and its constituents against Aedes albopictus', *Pest Management Science*, 71(11), pp. 1582–1586.
Manjunatha, M. *et al.* (1998) 'Response of western flower thrips, Frankliniella occidentalis and its predator Amblyseius cucumeris to chrysanthemum volatiles in olfactometer and greenhouse trials', *International Journal of Tropical Insect Science*, 18(2), pp. 139–144.
Manrique-Carpintero, N. C. *et al.* (2013) 'Sequence diversity in coding regions of candidate genes in the glycoalkaloid biosynthetic pathway of wild potato species', *G3: Genes, Genomes, Genetics*, 3(9), pp. 1467–1479.
Moore, G. (2015) 'Strategic pre-breeding for wheat improvement', *Nature Plants*, 1(3), pp. 1–3.
Ode, P. J. (2006) 'Plant chemistry and natural enemy fitness: effects on herbivore and natural enemy interactions', *Annual Review of Entomology*, 51, pp. 163–185.
Panda, N. and Khush, G. A. (1995) *Host plant resistance to insects*. CABI.
Peterson, C. J. *et al.* (1992) 'Genotype and environment effects on quality characteristics of hard red winter wheat', *Crop Science*, 32(1), pp. 98–103.
Powles, S. B. and Yu, Q. (2010) 'Evolution in action: plants resistant to herbicides', *Annual Review of Plant Biology*, 61, pp. 317–347.
Prescott-Allen, C. and Prescott-Allen, R. (1986) *The first resource*. Yale University Press.
Price, P. W. (1987) 'The role of natural enemies in insect', in *Insect outbreaks*, Editor: Bozzano G Luisa, Academic Press Inc, p. 287.
Ratnadass, A. *et al.* (2012) 'Plant species diversity for sustainable management of crop pests and diseases in agroecosystems: a review', *Agronomy for Sustainable Development*, 32(1), pp. 273–303.
Riahi, C. *et al.* (2022) 'Eliciting plant defenses through herbivore-induced plant volatiles' exposure in sweet peppers', *Frontiers in Ecology and Evolution*, p. 907.
Schoonhoven, L. M. *et al.* (2005) *Insect–plant biology*. Oxford University Press on Demand.
Shi-Guang, L. *et al.* (2017) 'Chemical composition of Illicium verum fruit extract and its bioactivity against the peach–potato aphid, Myzus persicae (Sulzer)', *Arthropod-Plant Interactions*, 11(2), pp. 203–212.
Singh, K. S. *et al.* (2021) 'Global patterns in genomic diversity underpinning the evolution of insecticide resistance in the aphid crop pest Myzus persicae', *Communications Biology*, 4(1), pp. 1–16.
Smith, C. M. (2005) *Plant resistance to arthropods: molecular and conventional approaches*. Springer Science & Business Media.
Snoeren, T. A. L. *et al.* (2010) 'The herbivore-induced plant volatile methyl salicylate negatively affects attraction of the parasitoid Diadegma semiclausum', *Journal of Chemical Ecology*, 36(5), pp. 479–489.
Sobhy, I. S. *et al.* (2020) 'Sensing the danger signals: cis-jasmone reduces aphid performance on potato and modulates the magnitude of released volatiles', *Frontiers in Ecology and Evolution*, 7(January), pp. 1–14. doi: 10.3389/fevo.2019.00499
Sutherland, O. R. W., Hutchins, R. F. N. and Wearing, C. H. (1974) 'The role of the hydrocarbon α-farnesene in the behaviour of codling moth larvae and adults', in *Experimental analysis of insect behaviour*, Editor: L. Barton-Browne, Springer, pp. 249–263.
Turlings, T. C. J. and Wäckers, F. (2004) 'Recruitment of predators and parasitoids by', in *Advances in insect chemical ecology*, Editors: Ring T. Cardé, Jocelyn G. Millar, p. 21.

Ulland, S. *et al.* (2008) 'Methyl salicylate, identified as primary odorant of a specific receptor neuron type, inhibits oviposition by the moth Mamestra brassicae L. (Lepidoptera, Noctuidae)', *Chemical Senses*, 33(1), pp. 35–46. doi: 10.1093/chemse/bjm061

van Emden, H. F. and Harrington, R. (2017) *Aphids as crop pests*. CABI.

van Emden, H. F. *et al.* (1969) 'The ecology of Myzus persicae', *Annual Review of Entomology*, 14(1), pp. 197–270.

Vet, L. E. M. and Dicke, M. (1992) 'Ecology of infochemical use by natural enemies in a tritrophic context', *Annual Review of Entomology*, 37(1), pp. 141–172.

Webster, B. *et al.* (2010) 'Volatiles functioning as host cues in a blend become nonhost cues when presented alone to the black bean aphid', *Animal Behaviour*, 79(2), pp. 451–457. doi: 10.1016/j.anbehav.2009.11.028

Wei, L. *et al.* (2014) 'Chemical composition and biological activity of star anise Illicium verum extracts against maize weevil, Sitophilus zeamais adults', *Journal of Insect Science*, 14(1), p. 80.

Wink, M. (1988) 'Plant breeding: importance of plant secondary metabolites for protection against pathogens and herbivores', *Theoretical and Applied Genetics*, 75(2), pp. 225–233.

Yan, F. *et al.* (2003) 'Roles of α-farnesene in the behaviors of codling moth females', *Zeitschrift für Naturforschung C*, 58(1–2), pp. 113–118.

Yang, G. *et al.* (2016) 'Electroantennogram and behavioral responses of Cotesia plutellae to plant volatiles', *Insect Science*, 23(2), pp. 245–252.

Zhao, J. *et al.* (2022) 'Evaluation and optimization of blends for attracting Trichogramma dendrolimi based on semiochemicals mediating tritrophic interactions in the orchard habitat', *Biological Control*, 173, p. 104998.

Zhu, X. *et al.* (2020) 'Chemical composition and attractant activity of volatiles from rhus potaninii to the spring aphid kaburagia rhusicola', *Molecules*, 25(15), p. 3412.

Züst, T. and Agrawal, A. A. (2016) 'Mechanisms and evolution of plant resistance to aphids', *Nature Plants*, 2(1), pp. 1–9.

7 Exogenous Application of *cis*-Jasmone Induces Defense in Brassica against *Myzus persicae*

7.1 INTRODUCTION

Plants have developed different defense mechanisms against herbivorous insects, including the release of a plethora of volatile organic compounds (VOCs) that directly repel the pests (Bruce and Pickett, 2011; Mithöfer and Boland, 2012). This implies that it might be possible to manipulate the emission of VOCs to enhance crop protection against insect pests (Turlings and Ton, 2006; Heil, 2014). By boosting signal transduction pathways for VOCs emission, qualitative and quantitative manipulation can be achieved (Thaler, Humphrey and Whiteman, 2012). The emission of VOCs can be increased and/or decreased by elicitors involved in these signalling pathways (Smith *et al.*, 2009). (Loughrin *et al.*, 1995; Birkett *et al.*, 2000; Röse and Tumlinson, 2004). Biosynthesis of methyl jasmonate, a catabolite of jasmonic acid, during herbivory activates defense mechanisms in tomato plants, *Lycopersicon esculentum* Mill. (Farmer and Ryan, 1990) while the exogenous spray of jasmonic and methyl jasmonate imitates the wound response similar to the attack of lepidopterous larvae (Tanaka *et al.*, 2009). In field conditions, the application of jasmonic acid to tomatoes is effective and made them more resistant against a number of pests (Tanaka *et al.*, 2009).

cis-Jasmone (CJ) is natural plant-derived compound which is biosynthesized via the isomerization of *cis*-oxophytodienoic acid (*cis*-OPDA) to iso-oxophytodienoic acid (iso-OPDA) following oxidative side-chain cleavage (Dabrowska and Boland, 2007). Factors such as herbivory (Loughrin *et al.*, 1995; Birkett *et al.*, 2000; Röse and Tumlinson, 2004), insect saliva application (Lou and Baldwin, 2003; Röse and Tumlinson, 2005; Sobhy, Erb and Turlings, 2015), spraying with jasmonic acid (JA) (Heil, 2004), or the introduction of nitrogen-fixing rhizobia (Ballhorn, Kautz and Schädler, 2013) can cause plants to release CJ (Loughrin *et al.*, 1995; Birkett *et al.*, 2000; Röse and Tumlinson, 2004). In addition to these factors, some plants also constitutively release CJ through flowers and leaves (Tanaka *et al.*, 2009), that could be used as a cue for host location by some herbivores, although the plant may produce it as a signal to attract pollinators (El-Sayed *et al.*, 2009).

CJ-induced plant defenses can be broadly categorized into direct and indirect defense; direct defense affects the performance of attacking herbivores including a reduced growth rate, lower fecundity (larviposition) and a higher mortality of pests

DOI: 10.1201/9781003400974-7

on CJ-treated plants (Sobhy *et al.*, 2020) while indirect defense shows an enhanced preference for natural enemies on CJ-treated plants (Bruce *et al.*, 2008). CJ-induced direct defense could be due to the production of antibiotic and defensive compounds, such as hydroxamic acid, benzoxazinoids and phenolic acid, that affect the physiological status of herbivorous pests (Pickett *et al.*, 2007a; Blassioli Moraes *et al.*, 2008) as well as causing the emission of plant volatiles that repel herbivores (Sobhy *et al.*, 2017, 2020). CJ treatment of plants alters tritrophic interactions between the host plant, herbivores and entomophagous insects because CJ-treated plants have an increased emission of volatile organic compounds (VOCs) that attract natural enemies (Oluwafemi *et al.*, 2013; Sobhy *et al.*, 2020). The CJ-induced plant volatile compounds belong to different classes of plant volatiles, including alcohols, aldehydes, aliphatic hydrocarbons, benzenoids, esters, ketones, nitrogen-containing compounds, and terpenes (Moraes *et al.*, 2008; Sobhy *et al.*, 2020). Many of these VOCs were found biologically active when tested in behavioral bioassays with insects such as *cis*-3-hexen-1-ol, methyl salicylate, caryophyllene and 4, 8, 12-trimethyl-(EE)-trideca-1, 37,11-tetraene (TMTT) (Bruce *et al.*, 2008). Previous studies showed that proved that the release of CJ can be used to induce plant defense (Birkett *et al.*, 2000), but these effects have not yet been studied in Brassica crops.

7.2 AIMS AND OBJECTIVES

The overall aim of this chapter is to assess the effect of CJ on brassica genotypes with regard to aphid performance on treated plants. Five *Brassica* lines, *B. napus* cv. 'Samurai', *B. napus* cv. 'Wesway', *B. napus* cv. English giant, *B. napus* cv. 'Turnip rutabaga 57' and *Brassica rapa* subsp. *chinensis* cv. Hanakan (Pak choi) were obtained from Warwick University, UK. Plants were sprayed with an aqueous emulsion of *cis*-Jasmone (CJ) (Sigma Aldrich, Buchs, Switzerland) as described in Bruce *et al.* (2003). Spray treatment was carried out using an Oshide spray bottle by applying three trigger pulls of spray formulation (250 µL) to each plant at a distance of 30 cm. Sprayed plants were left for 24 hours and then used for experiments. Control plants and CJ-treated plants were placed in different compartments to avoid any plant–plant signalling. Within this aim there are three objectives:

- To investigate the effect of CJ treatment of brassica crop on *M. persicae* performance
- To investigate the effect of CJ treatment of brassica crop on the performance of biocontrol agent (parasitoid) of *M. persicae*
- To investigate the effect on CJ treatment on volatile emission brassica lines

7.3 EFFECT OF CJ TREATMENT ON THE PERFORMANCE AND BEHAVIOR OF *MYZUS PERSICAE*

To investigate the effect of CJ treatment of brassica crop on the performance of *M. persicae*, a series of bioassays were performed, which are briefly described in this section:

7.3.1 Aphid Cage Bioassay

The performance of *M. persicae* was assessed on brassica genotypes. There were two separate series of experiments with different plants: the first series recorded observations after 48 hours and the second recorded observations after 96 hours. Fresh plants and aphids were used in each replicate observation in each experiment. In each replicate, 10 adult alate *M. persicae* were placed in a clip cage (2.5 cm diameter, Bioquip Products Inc. USA), which was attached to the lower surface of plant leaves (Sobhy *et al.*, 2020) (Figure 7.1). Two clip cages were placed on each plant. Ten replicates (control and CJ-treated) were performed for each genotype. To assess the survival and fecundity of aphids, plants were left undisturbed in a controlled environment room (25°C ± 2°C, 37% RH, 16 h L: 8 h D photoperiod). Plants were assessed after the 48-hour (series 1) or 96-hour (series 2) period. For assessment, leaves containing the cages were cut and cages were removed without losing any aphid. Numbers of live adults and newly produced nymphs were recorded.

After 48 h, there was no significant reduction in adult *M. persicae* survival on five brassica cultivars treated with CJ in clip-cage experiments (Figure 7.2a).The number of aphids surviving was lower on CJ-treated brassica cultivars but the difference was not statistically significant. A similar trend was observed after 96 hours, with no significant difference in aphid mortality was recorded on brassica cultivars except 'Samurai' (Figure 7.2b). Cultivar Samurai had an increase in number of dead aphids ($P = 0.04$) compared to control plants after 96 hours of CJ treatment.

There was a significant reduction in nymph production on CJ-treated plants across both time points (Figure 7.3a, b). Mean larviposition on CJ treated plants of all brassica cultivars was significantly reduced after 48 hours (Figure 7.3a), decreasing by 35% from 23.97 on control treated plants to 15.54 on CJ-treated plants. Larviposition was reduced most on Wesway ($P < 0.001$; 41% reduction in larviposition) and the

FIGURE 7.1 Clip-cage bioassay, two clip cages were placed on each plant, ten alate adult aphids were enclosed in each clip cage. Bioassays were run for 48 hours and 96 hours.

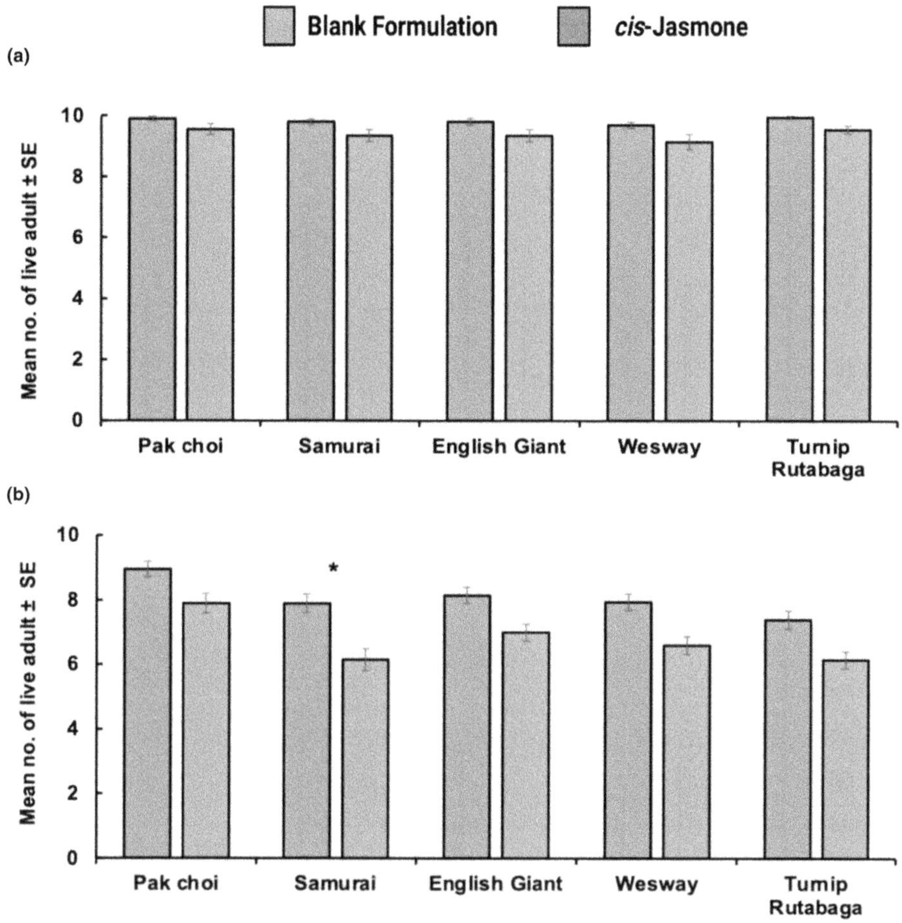

FIGURE 7.2 *Myzus persicae* adult survival (Mean ± SE) out of the original 10 individuals after (a) 48 hours and (b) 96 hours in clip cage on five brassica cultivars (n = 10) treated with *cis*-Jasmone or blank formulation (control). Asterisk denotes differing levels of statistical significance: * < 0.05, ** < 0.01 and *** < 0.001 (generalized linear models with Poisson probability distributions).

least on Samurai ($P < 0.001$; 21% reduction in larviposition). Similarly, mean larviposition on CJ-treated plants of all brassica cultivars was also significantly reduced after 96 hours (Figure 7.3b), decreasing by 39% from 57.18 on control plants to 34.81 on CJ-treated plants. Larviposition was reduced most on 'Samurai' ($P < 0.001$; 59% reduction in larviposition) and the least on 'Turnip Rutabaga' ($P = 0.004$; 24% reduction in larviposition).

7.3.2 Aphid Settlement Bioassay

Brassica plants were sprayed with *cis*-Jasmone and control plants sprayed with Tween 80 and tested under choice conditions in a Bug dorm-6E insect rearing cage (60 × 60 × 60 cm, 25°C ± 2°C, 16 h L: 8 h D photoperiod). Treatment was applied

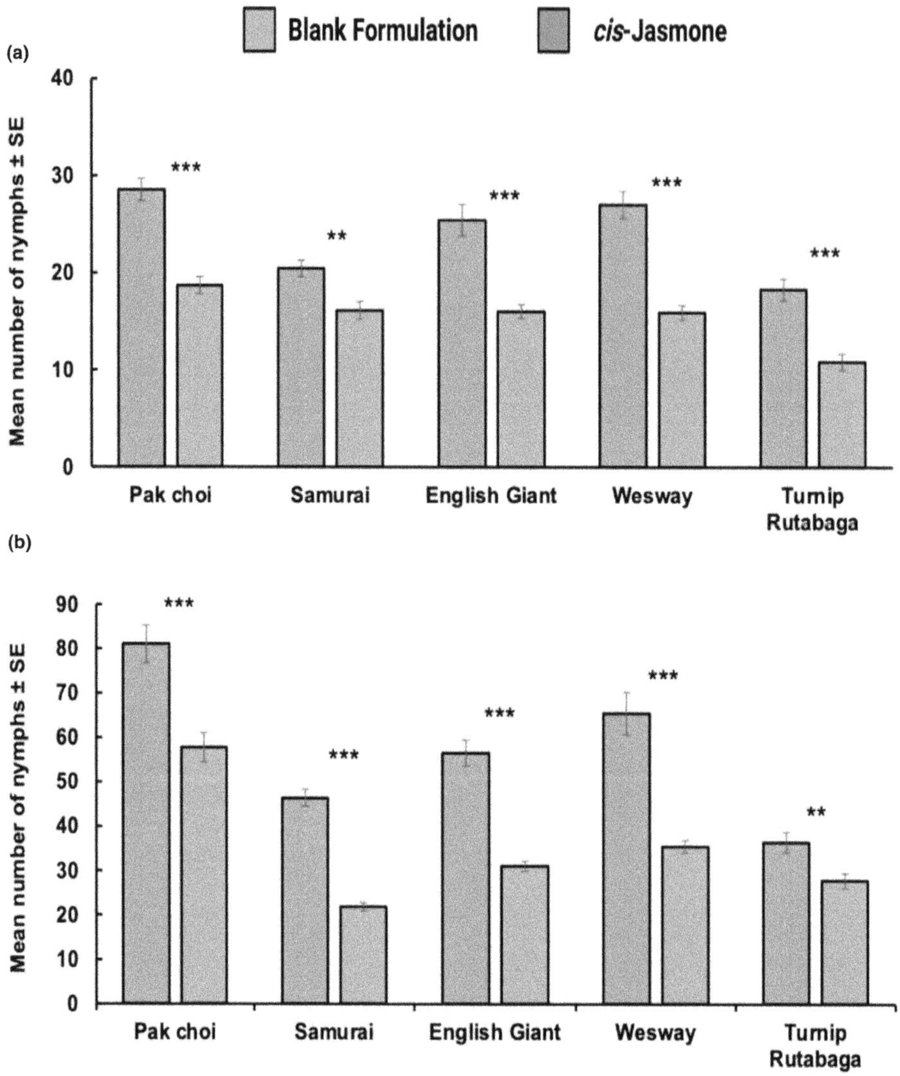

FIGURE 7.3 *Myzus persicae* larviposition (Mean ± SE) after (a) 48 hours and (b) 96 hours in clip cages on five brassica cultivars (n = 10) treated with *cis*-Jasmone or blank formulation (control). Asterisks denote differing levels of statistical significance: * < 0.05, ** < 0.01 and *** < 0.001 (generalised linear models with Poisson probability distributions).

24 hours before the start of the experiment. Four plants, two treated and two control, were placed inside the bug dorm at alternate positions (Figure 7.4). A vial (9 cm L × 2.5 cm diameter) containing 50 alate *M. persicae* was placed in the bug dorm, opened and then the bug dorm was left undisturbed for the next 24 hours. Counts of settled aphids were made after 24 hours after release, and settlement on treated plants was compared with settlement on control plants using a t-test.

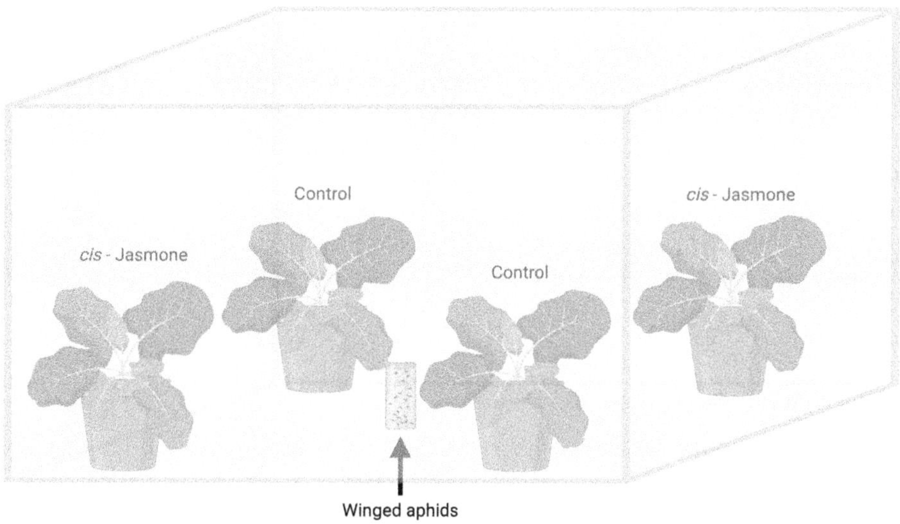

FIGURE 7.4 Settlement bioassay; two control and two CJ-treated plants were placed in a Bugdorm as shown in the figure. A vial containing 50 winged aphids was placed in the middle.

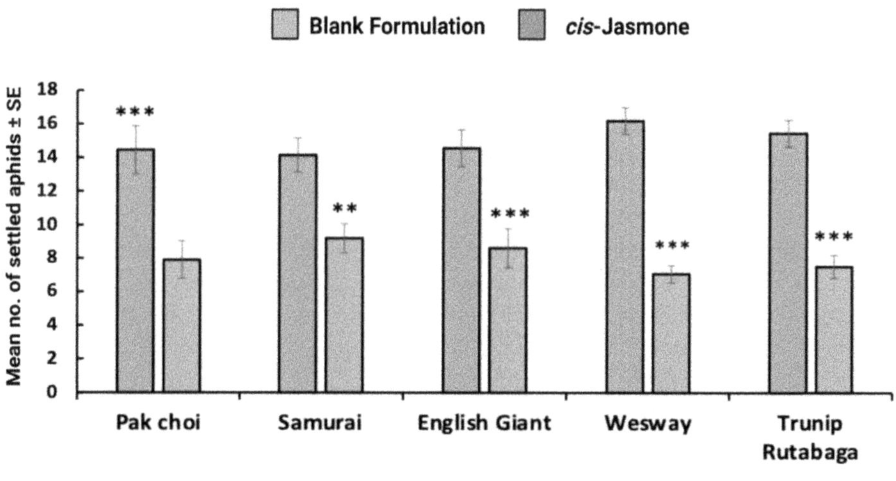

FIGURE 7.5 Mean settlement of *Myzus persicae* after 24 hours on untreated (control) and *cis*-Jasmone (CJ)-treated plants in a choice bioassay (50 winged aphids were released in each replicate $n = 10$). Error bars represent standard error of the mean (± SE). Asterisks denote different levels of statistical significance * < 0.05, ** < 0.01 and *** < 0.001 (generalised linear models with either Poisson or quasi-Poisson probability distributions).

In settlement bioassay, where aphids were offered a choice between CJ-treated and control plants, a clear and statistically significant reduction in aphid settlement was observed on CJ-treated plants (Figure 7.5). This effect was consistent across all brassica cultivars tested. The preference for control over CJ-treated plants was strongest for the *B. napus* cultivar Wesway, with a mean of only 7.05 aphids settling on CJ-treated plants compared to 16.2 aphids settling on control plants (GLM with quasi-Poisson distribution: $F = 32.39$; *d.f.* = 1,38; $P < 0.001$; 30.3% of aphid settlement occurring on CJ-treated plants). Similarly, aphid settlement was significantly reduced in Pak choi (GLM with quasi-Poisson distribution: $F = 98.81$; *d.f.* = 1,38; $P < 0.001$), English Giant (GLM with Poisson distribution: $X2 = 86.63$; *d.f.* = 1,38; $P < 0.001$), Samurai (generalized linear model with Poisson distribution: $X2 = 70.4$: *d.f.* = 1,38; $P < 0.001$), and Turnip Rutabaga (GLM with Poisson distribution: $X2 = 41.12$; *d.f.* = 1,38; $P < 0.001$). Pooling data across all cultivars tested, mean aphid settlement on CJ-treated plants was 1.86 times lower than on control plants.

7.3.3 Aphid Behavioral Responses

A Perspex four-arm olfactometer (Pettersson, 1970) was used to determine the behavioral responses of *M. persicae* to host volatiles in a controlled environment room (24°C ± 2°C, 30% RH). Prior to each experiment, all Perspex components were washed with Teepol solution and then rinsed with 80% ethanol solution and distilled water and left to air dry. Filter paper (Whatman No. 1, Buckinghamshire, UK) was fitted at the base of the olfactometer to provide purchase for the walking insects. Air was drawn out through the central hole on top of the olfactometer by PTFE tubing. A single aphid was introduced through a central hole at the top of the olfactometer. The air was drawn out through the central hole at a rate of 200 mL/min. Aphids were acclimatized in the room for 2 hours, and each replicate of the experiment was run for 12 min. After every 3 min., the olfactometer was rotated 90° to avoid any directional bias in the experiment. Volatile samples were collected from brassica plant lines that were used as odor sources. Odor sources were prepared by putting an aliquot (10 μL) of sample on the filter paper, inserted into odor source side arms. The olfactometer was divided into five regions. Regions 1 to 4 corresponded to each of the four plastic side-arms containing odor sources and region 5 was the central region. A filter paper strip containing aliquot of plant odor source was inserted in arm 1 while the rest of the three arms contained control sample (diethyl ether). Time spent in each region was recorded using Olfa software (F. Nazzi, Udine, Italy). If an aphid remained inactive (motionless) for the first 2 minutes of a replicate that replicate was discarded. Ten replicates were performed for each odor source. All bioassays were performed between 10:00 and 13:00.

When presented with a choice between a solvent control (diethyl ether; DEE) and volatiles collected from plants treated with a blank formulation, adult aphids showed no significant preference for either odor source in five of the tested cultivars (Figure 7.6a). Aphids did, however, spend significantly less time in the treated zone for Wesway (paired *t*-test: $t = 4.69$; *d.f.* = 9; $P = 0.001$) (Figure 7.6a). However, when plants were CJ-treated and aphids were presented with a choice of volatiles from a treated plant versus solvent control (DEE), aphids spent significantly less time in the treated zone of the olfactometer with volatiles from CJ-treated plants for 4 out of the 5 cultivars tested (Figure 7.6b). 'Turnip Rutabaga' was the only cultivar in which volatiles from CJ-treated plants were not repellent. CJ-treated 'Samurai' and 'Pak choi' were the most repellent. When also allowed to choose between volatiles from

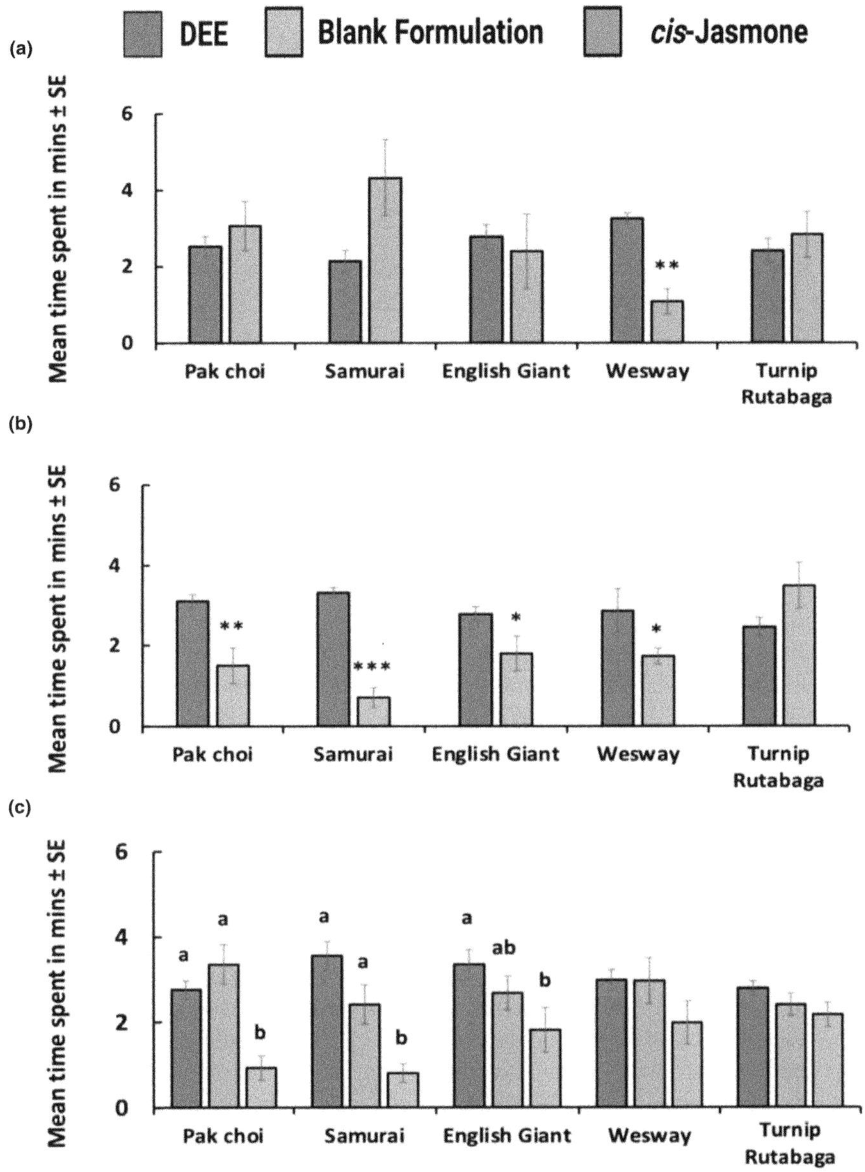

FIGURE 7.6 Behavioral responses of *Myzus persicae* females to volatile organic compounds (VOCs) from five brassica genotypes in an olfactometer bioassay. Individual aphids were given 12 minutes to make a choice between (a) one arm of blank formulation-treated plants (BF – tween 80 and water) vs. three solvent diethyl ether (DEE) arms, (b) one arm of *cis*-Jasmone treated plants (CJ) vs. three solvent DEE arms and (c) two different treatment arms (BF = blank formulation-treated plants and CJ = *cis*-Jasmone-treated plants) vs. two solvent DEE arms. The values shown are mean time spent in arm ± standard error (SE) ($n = 10$). For Figure 7.6a and b, Brassica genotypes capped with asterisks denote differing levels of statistical significance: * < 0.05, ** < 0.01 and *** < 0.001 (paired *t*-test). For Figure 7.6c, different letters above bars indicate statistically significant differences between treatments ($P < 0.05$), based on post-hoc tests.

CJ-treated plants and untreated control plants (Figure 7.6c), aphids spent significant longer in the olfactometer zone with volatiles from blank formulation control plants for 'Pak choi' and 'Samurai' cultivars.

7.3.4 Discussion

The current study supports previous studies showing that *cis*-Jasmone treatment of plants made them more resistant to insect pests (Bruce *et al.*, 2003; Sobhy *et al.*, 2017, 2020). Various crop plants, including tomato, maize, wheat potato, soybean treated with CJ, were shown to have enhanced levels of resistance (Bruce *et al.*, 2008; Moraes *et al.*, 2009; Oluwafemi *et al.*, 2013; Disi *et al.*, 2017; Sobhy *et al.*, 2017). CJ application activates the plant defense system to produce volatile organic compounds that have a repellent effect on pests (Birkett *et al.*, 2000; Bruce *et al.*, 2008; Oluwafemi *et al.*, 2013). In this study, several bioassays were performed to assess the performance of *M. persicae* on CJ-treated and control brassica crop plants that had not been tested previously. We found that the performance of *M. persicae* was reduced on treated plants in terms of survival and fecundity. *M. persicae* also showed less preference for CJ-treated plants than untreated control plants.

All five brassica cultivars did not show any significant difference in aphid mortality in clip-cage bioassay after 48 hours. However, the number of live adults was lower on CJ-treated plants compared to control plants. The effects of CJ treatment of brassica cultivars on aphids were more apparent after 96 hours compared to 48 hours since the aphids have been on the plants for longer time and therefore were more affected by any negative effects of the CJ treatment. Samurai was the most responsive cultivar with a significant high increase in the mortality of aphids on CJ plants. Overall, the number of live aphids was lower on all CJ plants throughout all five cultivars. In contrast to adult aphids, CJ treatment had a strong effect on nymph production; as per results, it can be seen that plants treated with CJ were less suitable for aphid reproduction. Data collected from both series (48 hours and 96 hours) showed that CJ treatment induces changes in plant physiology that affect the aphid's reproductive capability. If the experiment was continued for longer than 96 hours it is likely that even bigger effects would have been observed. The combined effect of increased mortality and reduced larviposition can be considered to 'slow down' aphid population growth. Lower reproductive rate is an important effect of CJ treatment because the aphid's fast reproductive rate is one of the main characteristics that makes it one of the most notorious pests (Leather, 2017). The difference in aphid survival and larviposition is due to the upregulation of plant defense induced by CJ treatment (Dewhirst *et al.*, 2012; Sobhy *et al.*, 2020).

Settlement bioassays showed that aphid colonization was significantly lower on CJ-treated plants for all five cultivars. This experiment shows that low aphid colonization would further slow down the aphid infestation, which would be an important step in controlling the aphid population on the plant. Behavioral bioassays revealed that aphids spent less time in the arm of olfactometer treated with CJ treated volatiles i.e. CJ-treated plants were repellent to aphids. In olfactometer bioassays, volatiles collected from CJ-treated plants were repellent to aphids for most cultivars. Surprisingly, volatiles collected from Turnip rutabaga showed no repellent effect on aphids, although volatile analysis showed a fourfold increase in volatile emission with CJ treatment. This suggests that the composition or quality of volatile components is more important than the quantity of the component to repel the pest. In addition, a repellent effect was observed in another brassica cultivar (Chinese kale), although volatile emission was not increased by CJ treatment (Ali *et al.*, 2021). This again proves the type of volatile chemicals released is important. Based on these results we can say CJ acts as a plant defense elicitor to induce brassica crop plant resistance against aphids.

The current results showed a similar pattern to previously reported results for CJ-induced defense with other plant species. For instance, CJ-treated plants showed reduced pest growth and reproduction, reduced settlement of the pest population and volatiles collected from CJ plants were avoided. These effects have been observed in several other crop plants, such as wheat, potato and soybean (Bruce *et al.*, 2003; da Graça *et al.*, 2016; Bayram and Tonğa, 2018; Sobhy *et al.*, 2020) but had not been tested in brassica crops previously. To test the effect of CJ on aphid natural enemies, we have performed further bioassays in the following chapter.

7.4 EFFECT OF CJ TREATMENT ON PARASITOID PERFORMANCE

To investigate the effect of CJ treatment of brassica crop on the performance of parasitoid *D. rapae*, a series of bioassays were performed, which are briefly described in the following section.

7.4.1 Parasitoid Foraging Bioassay

All experiments were conducted under the same conditions of temperature and daylength as used for insect rearing. CJ-treated and control plants were used in the experiments. Each line was replicated ten times for CJ-treated and control. Treated and control replicates were observed alternately. All replicates done in a single day were of the same line. All experiments were conducted in an open-fronted cage. For each replicate, the plant was placed on a turntable at the centre of the cage. With the help of a vial a single female parasitoid was collected from the Bugdorm and released onto the leaf of the plant and its foraging behavior was monitored by direct observation, until the parasitoid left the plant. Noldus Observer 4.1 software was used to record the behavioral observations. All parasitoids used were between 24 hours and 32 hours old and each used only once and then discarded. All experiments were done between 9:00 hours and 16:00 hours. The following data were recorded during observations of parasitoid foraging: (i) type of behavior (walking, still, cleaning and flying); (ii) duration of each period of walking, still, cleaning and flying (seconds); (iii) total time spent on the plant by each parasitoid (seconds). Time spent walking, still, and cleaning was recorded, as well as the total time spent before the parasitoid left the plant. An observation was terminated when a parasitoid flew away from the plant, which was considered the foraging "patch." Time spent on treated and control plants was compared by using a paired *t*-test.

In a foraging bioassay, parasitoid wasps spent substantially longer on CJ-treated plants than on control plants (Figure 7.7). There was a 5.1x increase in mean time

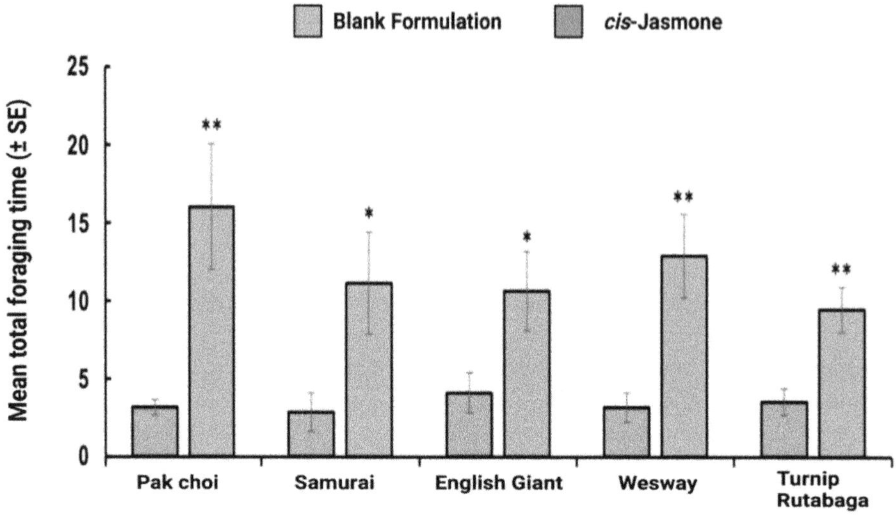

FIGURE 7.7 Mean total time spent foraging (± SE) by *Diaeretiella rapae* on untreated (control) and *cis*-Jasmone (CJ)-treated plants ($n = 10$). Error bars represent standard error of the mean (SE). Brassica genotypes capped with asterisks show a significant difference between control and CJ treatment with differing levels of statistical significance: * < 0.05, ** < 0.01 and *** < 0.001 (two-sample *t*-tests).

spent on CJ-treated Pak choi plants (two-sample *t*-test, $P = 0.004$), a 4.6x increase on Turnip Rutabaga (two-sample *t*-test, $P = 0.001$), a 4.5x increase on Wesway (two-sample *t*-test, $P = 0.001$), a 3.9x increase on Samurai (two-sample *t*-test, $P = 0.013$), and a 2.8x increase on English Giant (two-sample *t*-test, $P = 0.04$).

7.4.2 Parasitism Bioassay

In this choice test bioassay, the effect of CJ on the parasitism rate of *D. rapae* was assessed on CJ-treated and control plants in a Bugdorm cage (60 × 60 × 60 cm; NHS Ltd), which was kept in a controlled environment room (25°C ± 2°C, 37% RH, 16 h L: 8 h D photoperiod). Each Bugdorm contained four plants (two CJ-treated and two control) and the plants were placed at alternate positions (Figure 7.8). Each plant was inoculated with 50 adult aphids 2 hours prior to the release of parasitoids *D. rapae*. Eight female parasitoids were released in the Bugdorm using a plastic vial following the experimental procedure described in Sun *et al.* (2020). Parasitoids were released to observe the difference in parasitism behavior of *D. rapae* on CJ and control plants infested with aphids. After 24 hours, parasitoid females were collected from the cages. Experiments were conducted in a climate-controlled room (25°C ± 2°C, 37% RH, 16 h L: 8 h D photoperiod). After 15 days, the number of mummies collected from treated and control plants was recorded. The assays were repeated ten times on different experimental days.

In parasitism bioassay, three out of five CJ-treated brassica lines showed a significant increase in the number of parasitized aphids (mummies) (Figure 7.9). The largest

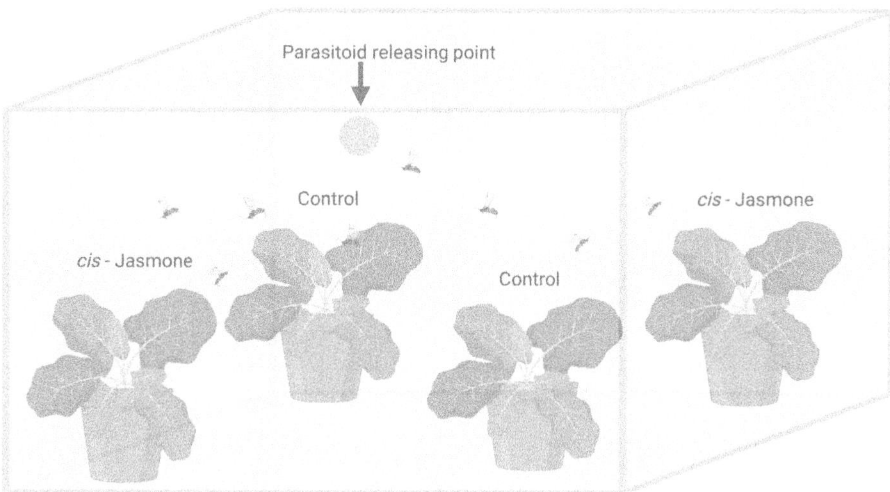

FIGURE 7.8 Experimental setup for assessing parasitism behavior of female *D. rapae* on aphid-infested (control vs. CJ) plants. Each plant was infested with 50 adult aphids 2 hours prior to release of the parasitoid. Plants were removed after 24-hour exposure to parasitoids in the Bugdorm cage.

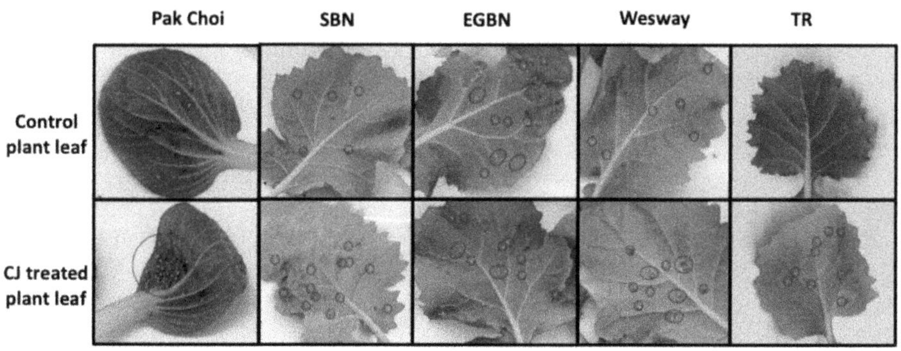

FIGURE 7.9 Brassica plants leaves (Control and CJ) showing number of mummified aphids (red circle).

increase in parasitism was observed on Samurai ($P = 0.001$; 121% increase), though Pak choi had the greatest total number of mummified aphids. (Figure 7.10).

7.4.3 Parasitoid Behavioral Responses

Olfactometer bioassay series 1 was a choice between a solvent control (diethyl ether; DEE) and volatiles collected from plants treated with a blank formulation. Genotypes Wesway and English Giant showed significant deterrence to parasitoids ($P = 0.0001$ and 0.002 respectively) (Figure 7.11a). Parasitoids spent more time in the treated zone for the other three genotypes, but the difference was not significant (Figure 7.11a).

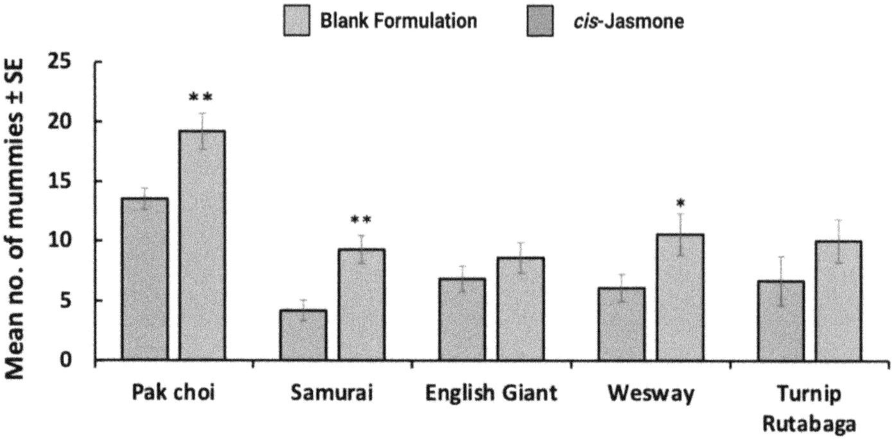

FIGURE 7.10 Mean number (± SE) of mummified *M. persicae* on blank formulation and *cis*-Jasmone treated plants 15 days after exposure to parasitoid *D. rapae* ($n = 10$). Brassica cultivars capped with asterisks show differing levels of statistical significance: * < 0.05, ** < 0.01 and *** < 0.001 (two-sample *t*-tests).

However, in series 2, when plants were CJ-treated and parasitoids were presented with a choice of volatiles from a treated plant versus solvent control (DEE), parasitoids spent significantly more time in the treated zone of the olfactometer with volatiles from CJ-treated plants for 3 out of the 5 cultivars tested (Figure 7.11b). English Giant had the highest parasitoid preference for volatiles from CJ-treated plants ($P < 0.01$) while Samurai had the least preference ($P > 0.05$). In series 3, when parasitoids were also allowed to choose between volatiles from CJ-treated plants and untreated control plants (Figure 7.11c), parasitoids spent significantly longer in the olfactometer zone treated with volatiles from CJ plants for all brassica cultivars tested: Pak choi ($P < 0.04$), Samurai ($P < 0.01$), Turnip rutabaga $P < 0.001$, Wesway ($P < 0.001$) and English Giant ($P = 0.003$). An increase of 291% and 64% in mean time spent by *D. rapae* was recorded when the olfactometer arm treated with CJ-treated Turnip rutabaga and Samurai plant volatiles was compared to the arm treated with plant volatiles treated with blank formulation respectively.

7.4.4 Discussion

This section describes the effect of *cis*-Jasmone treatment of brassica crop plants on the behavior of aphid parasitoid *D. rapae*. A foraging bioassay was performed to observe the time duration parasitoids spent on blank formulation and CJ-treated plants. This bioassay was performed because retention for foraging after arrival is important for a good level of parasitism in the field (Budenberg, Powell and Clark, 1992). Similar studies, performed with the model plant *Arabidopsis*, showed that parasitoids spent double the time on CJ-treated plants compared to control plants (Bruce *et al.*, 2008). Longer retention time on plants increases the opportunity for the parasitoid to parasitize the pest. Here, we also found a significant increase in foraging

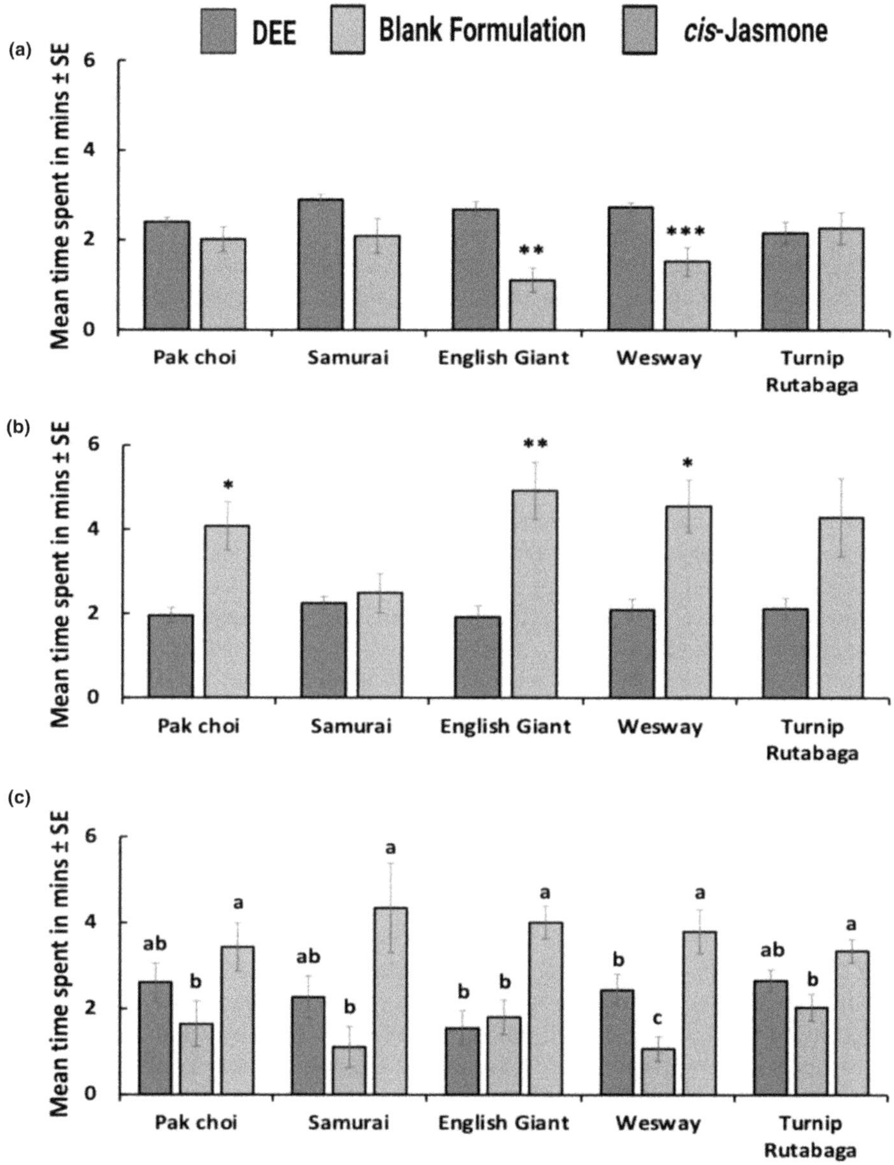

FIGURE 7.11 Behavioral responses of *Diaeretiella rapae* females to volatile organic compounds (VOCs) from five brassica genotypes in an olfactometer bioassay. Individual parasitoids were given 12 minutes to make a choice between: (a) one arm of blank formulation-treated plants (BF – tween 80 and water) vs. three solvent diethyl ether (DEE) arms; (b) one arm of *cis*-Jasmone-treated plants (CJ) vs. three solvent DEE arms; and (c) two different treatment arms (BF = blank formulation treated plants and CJ = *cis*-Jasmone-treated plants) vs. two solvent DEE arms. The values shown are mean time spent in arm ± SE ($n = 10$). For Figure 7.11a and b, Brassica genotypes capped with asterisks denote differing levels of statistical significance: * < 0.05, ** < 0.01 and *** < 0.001 (paired *t*-test). For Figure 7.11c, different letters above bars indicate statistically significant differences between treatments ($P < 0.05$), based on post-hoc tests.

time, with parasitoids spending up to 4.5 times more time on CJ-treated plants than on control plants. The increase in foraging time observed on the brassica cultivars in the current study could be because of changes induced by CJ treatment in host plants as reported in previous studies (Moraes *et al.*, 2008; Bruce *et al.*, 2008; Bayram and Tonğa, 2018). The effect of CJ treatment on the volatile profile of brassica plants will be discussed in the next section. The change in volatile emissions of treated plants could influence the foraging behavior of parasitoids.

In addition, another experiment 'parasitism bioassay' was also performed that supported the findings of the foraging bioassay. An overall increase of 50% in number of mummified aphids was recorded on CJ-treated plants when compared with BF-treated plants. Out of five brassica cultivars, three cultivars showed a significantly higher number of mummified aphids. In particular, cultivar Samurai, Wesway and Pak choi showed a significant increase of 2.21x, 1.53x and 1.42x fold, respectively, in the number of mummified aphids on CJ plants compared to control plants, while Turnip rutabaga and English Giant showed the least increase. The effect of CJ application at the tritrophic level has been tested with several plants, and most of these previous studies support its role as a stimulator of biological control agents. The behavioral responses of several control agents have been observed with *cis*-Jasmone compound or CJ-treated plants; for instance, *Coccinella septempunctata* showed a preference for CJ in a wind tunnel bioassay when exposed to CJ-induced bean plant volatiles (Birkett *et al.*, 2000),

To investigate the behavioral response of *D. rapae* to plant volatiles a series of olfactometer bioassays was performed using different combinations of solvent blank control and plant volatile treatments collected from blank formulation and CJ-treated plants. Parasitoids spent a longer time in the zone treated with volatiles collected from CJ-treated plants. Plant volatiles collected from CJ-treated plants had a significant attractant effect on *D. rapae*. All five genotypes had CJ-induced plant volatiles that acted as an attractant for parasitoids. Cultivars 'Wesway' and 'English Giant' showed the highest level of induced effect after CJ treatment and were highly attractive to *D. rapae*. As mentioned in the last section, olfactometer studies again proved that changes in the plant volatile blend can play a critical role in inducing insect behavioral changes. The high preference of *D. rape* to CJ plants may be due to the induced effect of CJ in brassica lines. For instance, English Giant and Wesway volatile samples were repellent before CJ treatment but after treatment volatiles samples were highly attractive to *D. rapae*.

An earlier study on Arabidopsis reported that CJ-treated Arabidopsis plants also release volatile compounds that act as an attractant to *Aphidius ervi* which spent a longer time foraging on CJ plants compared to control (Bruce *et al.*, 2008). In tobacco, the parasitic efficiency of *Campoletis chloridae* was enhanced when it was exposed to *Helicoverpa armigera*-infested plants treated with CJ (Sun *et al.*, 2020). *Telenomus podisi* was highly attracted to soybean plants exposed to CJ application (Moraes *et al.*, 2009). During a field experiment, CJ treatment of wheat plants was responsible for the attraction of aphid parasitoids; *Collyria coxator* and predator ladybird beetle *Coccinella septempunctata* (Bayram and Tonğa, 2018). The current findings suggest that CJ treatment could be used to induce plant defense that will improve the biological control of aphids in brassica crops by increasing parasitoid visits and foraging time. To investigate the differences induced by CJ in volatile emission, we have performed volatile analysis in the following section.

7.5 EFFECT OF CJ TREATMENT ON VOLATILE EMISSION

Any qualitative or quantitative change in plant volatile emission can affect plant–herbivore interactions (Silva *et al.*, 2017; Takabayashi and Shiojiri, 2019). This change can be the consequence of any abiotic (light, drought, salinity, nutrients) and/or biotic stresses (pest and pathogens) to which the plant is exposed (Gouinguené and Turlings, 2002; Pedrol, González and Reigosa, 2006). During biotic stress, the plant recognises the damage caused and enzymes released into the host body by the invaders, and activates defense pathways that result in the production of antibiotic and defensive organic compounds that repel the stress-causing entity (Giron *et al.*, 2016; Wielkopolan and Obrępalska-Stęplowska, 2016; Ighodaro and Akinloye, 2018; Turlings and Erb, 2018).

In Sections 7.3 and 7.4, performance and behavioral responses of *M. persicae* and *D. rapae* to the odor of the CJ-treated brassica lines were studied. *M. persicae* and *D. rapae* respond to plant odor by spending less and more time in the treated zone of odor collected from CJ-treated plants respectively, indicating the repulsion of aphids but the attraction of parasitoids. This behavioral response of insects may be due to the presence of a single volatile compound or a blend of more than one compound. Plants produce a plethora of volatile compounds and so the first step is to collect and analyse an entrainment sample for volatile analysis to get information of the compounds present in the sample. Using coupled gas chromatography-mass spectrometry (GC-MS), it is possible to analyse and identify the compounds present in the volatile blend. This chapter provides details of work carried out in analysing the volatile compounds collected from CJ-treated and untreated control brassica plants by coupled GC-MS.

7.5.1 Volatile Analysis

GC-MS analysis of volatile samples collected from five cultivars of brassica found 24 detectable volatile compounds belonging to 8 functional classes (alcohols, aldehydes, aliphatic hydrocarbons, benzenoids, esters, ketones, nitrogen-containing compounds and terpenes). CJ application induced qualitative and quantitative changes in volatile profiles. CJ-treated plants showed an increase in volatile emissions. All five cultivars had quantitative changes in volatile profiles: volatile emission was increased the most in Wesway (up to 14-fold), while the lowest increase was recorded with English Giant (2-fold) (Figure 7.12). Table 7.1 shows the details of the compounds identified from control and treated plants. CJ-treated English Giant produced six main compounds; nonanal, (*E*)-3-tetradecene, dihydrojasmone, CJ, methyl isothiocyanate and benzyl nitrile, while Pak choi showed a large increase in green leaf volatile (*Z*)-3-hexenyl acetate and 2-ethyl-1-hexanol production together with CJ, limonene and citronellol after treatment with CJ. Samurai was the brassica cultivar that showed a high emission of CJ itself, together with a smaller increase in the emission of (*Z*)-3-hexenyl acetate and 2-ethyl-1-hexanol and β-elemene. β-Elemene,

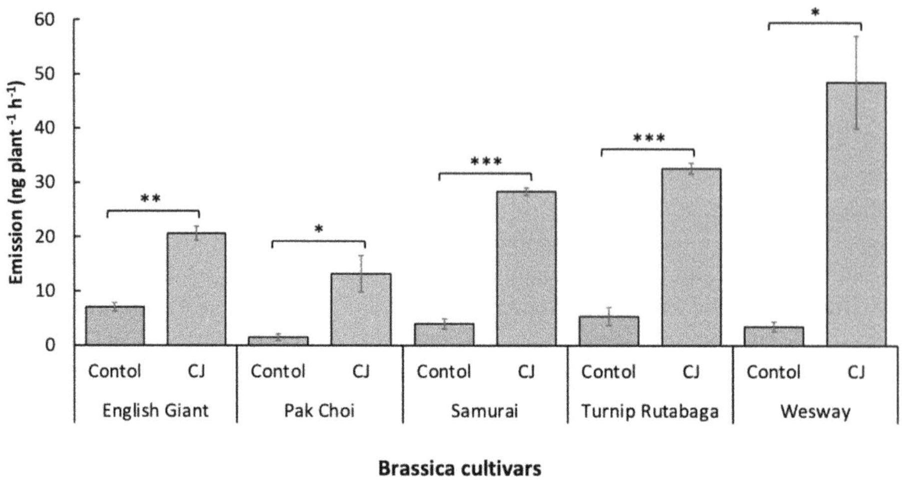

FIGURE 7.12 Total amount (mean nanogram plant $^{-1}$ h^{-1} ± SE) of identified volatile organic compounds (VOCs) emitted from the five brassica cultivars with and without CJ treatment. Asterisks indicate statistically significant differences: *** < 0.001 and * < 0.005 (paired *t*-test).

(*E*,*E*)-α-Farnesene, 2-ethyl-1-hexanol and p-Cymen-7-ol with the notable induction of methyl salicylate (MeSA) and CJ were the main compounds release by turnip rutabaga. A marked increase in emission of about 10 volatiles, including CJ, 2-ethyl-1-hexanol, (*Z*)-3-hexenyl acetate, β-elemene and p-Cymen-7-ol, was recorded in Wesway. The quantitative changes in volatile emission are shown in Figure 7.12.

7.5.2 Behavioral Response of Aphids and Parasitoids to the Synthetic Analogue of Identified Compounds

A four-arm olfactometer was used to determine the behavioral response of *M. persicae* and *D. rapae* to the volatile compounds as described in Section 7.3.3. An aliquot (10 μL) of each test compound of standard concentration (100 ng/μL) was applied to filter paper strips. Ten μL hexane was applied on filter paper strips as solvent (control). Insects were exposed to synthetic volatile compounds for 12 minutes and ten replicates were performed for each compound. The behavioral responses of *M. persicae* and *D. rapae* were tested to the synthetic analogues of compounds identified from headspace samples collected from five brassica cultivars.

The results obtained from behavioral bioassay using synthetic analogues of identified compounds (Figure 7.13a) shows time spent in the treated region of olfactometer compared to control when each compound was tested individually at the standard concentration of 100 ng. Of 23 volatile compounds, six compounds elicited a significant change in aphid behavior. Of these six compounds, five were repellent to aphids while one acted as an attractant. The compounds that had a repellent effect were methyl isothiocyanate ($p = 2.536$E-05), *cis*-Jasmone ($p = 0.03$), β-elemene ($p = 0.006$), (*E*)-3-tetradecene ($p = 0.04$), and methyl salicylate ($p = 0.03$). Among these six compounds, methyl isothiocyanate and β-elemene were the most repellent. In contrast,

TABLE 7.1
Emission (in ng; mean ± SE; $n = 3$) of Volatiles Released by *cis*-Jasmone Treated (CJ) and Blank Formulation Treated (Control) Plants[1]

Plant volatile	KI	English Giant		Pak Choi		Samurai		Turnip Rutabaga		Wesway	
		Control	CJ	Control	CJ	Control	CJ	Control	CJ	Control	CJ
Alcohols											
2-ethyl-1-hexanol	1030	0.6±0.1	0.7±0.1	**0.34±0.02**	**2.17±0.3**	**0.51±0.06**	**3.35±0.4**	**0.97±0.5**	**3.61±0.5**	**0.36±0.2**	**5.92±0.9**
1-octanol	1071	ND	ND	ND	ND	**ND**	**0.31±0.08**	**ND**	**0.37±0.1**	**ND**	**0.67±0.2**
2-butyl-1-octanol	1277	0.02±0.01	0.6±0.2	**ND**	**0.59±0.2**	0.06±0.04	0.41±0.2	**0.03±0.02**	**0.52±0.1**	0.05±0.02	1.41±0.5
Aldehydes											
Nonanal	1104	**0.7±0.2**	**3.4±0.4**	ND	0.06±0.04	**0.07±0.06**	**1.15±0.2**	**0.43±0.2**	**1.61±0.3**	**0.08±0.06**	**2.62±0.6**
Decanal	1206	**0.03±0.01**	**0.17±0.01**	ND	0.17±0.06	0.05±0.04	0.34±0.1	0.22±0.1	0.41±0.02	**0.04±0.03**	**0.97±0.2**
Aliphatic hydrocarbons											
Dodecane	1200	0.06±0.05	0.13±0.03	0.03±0.01	0.07±0.05	0.04±0.03	0.17±0.09	0.05±0.04	0.18±0.07	**ND**	**0.62±0.06**
(*E*)-3-tetradecene	1385	0.4±0.2	1.5±0.2	1.16±0.1	1.34±0.6	0.18±0.07	0.92±0.6	0.13±0.1	0.51±0.2	**ND**	**3.48±0.9**
Benzenoids											
MeSA[#]	1192	ND	ND	ND	ND	ND	ND	**0.05±0.04**	1.57±0.4	ND	2.5±0.6
Benzothiazole	1229	ND	ND	ND	0.19±0.1	ND	0.77±0.3	**ND**	**0.63±0.1**	ND	0.39±0.3
Esters											
cis-3-hexenyl acetate	1005	2.3±0.3	2.9±0.5	0.43±0.3	2.81±1.2	**0.91±0.1**	**2.32±0.2**	**1.05±0.2**	**2.11±0.09**	**1.28±0.4**	**5.3±0.8**
2-ethylhexyl acetate	1129	0.17±0.01	0.22±0.01	0.08±0.03	ND	0.14±0.07	0.19±0.08	0.05±0.03	0.19±0.01	ND	ND
Ketones											
Dihydrojasmone	1369	0.10±0.04	1.09±0.6	**ND**	**0.27±0.03**	**ND**	**0.63±0.07**	ND	ND	ND	ND
cis-Jasmone	1394	**0.12±0.06**	**1.53±0.3**	**ND**	**1.76±0.5**	**ND**	**10.9±2.7**	**0.16±0.13**	**1.6±0.4**	**ND**	**11.7±2.9**

(Continued)

TABLE 7.1 (CONTINUED)

Emission (in ng; mean ± SE; n = 3) of Volatiles Released by *cis*-Jasmone Treated (CJ) and Blank Formulation Treated (Control) Plants[1]

Plant volatile	KI	English Giant		Pak Choi		Samurai		Turnip Rutabaga		Wesway	
		Control	CJ	Control	CJ	Control	CJ	Control	CJ	Control	CJ
N-containing compounds											
Methyl isothiocyanate	992	**0.08±0.06**	**0.5±0.1**	0.05±0.04	0.09±0.07	**0.09±0.07**	**0.42±0.01**	ND	ND	ND	0.23±0.2
Benzyl nitrile	1144	**ND**	**1.7±0.2**	0.04±0.03	0.73±0.24	ND	ND	**ND**	**1.26±0.3**	ND	0.95±0.8
Terpenes											
D-limonene	1030	0.2±0.1	0.5±0.1	ND	1.97±1.1	0.61±0.4	0.56±0.2	0.41±0.2	1.27±0.4	ND	2.17±1.1
Eucalyptol	1032	ND	0.3±0.1	ND	ND	0.07±0.05	ND	ND	ND	ND	ND
Citronellol	1229	0.15±0.09	0.22±0.1	**ND**	**1.03±0.2**	0.11±0.08	ND	ND	ND	ND	ND
DMNT#	1116	ND	ND	ND	ND	ND	ND	**ND**	**0.79±0.2**	ND	ND
p-cymen-7-ol	1289	0.35±0.2	0.54±0.08	**0.24±0.12**	**0.85±0.02**	0.32±0.08	0.67±0.3	**0.32±0.1**	**3.31±0.5**	**0.23±0.1**	**4.67±1.2**
α-cedrene	1411	ND	ND	0.06±0.02	0.24±0.05	ND	ND	ND	ND	ND	ND
β-elemene	1391	**1.41±0.1**	**2.47±0.1**	ND	ND	**0.75±0.1**	**2.79±0.4**	**0.47±0.3**	**7.7±1.7**	**1.23±0.4**	**4.75±0.5**
β-curcumene	1514	0.18±0.02	0.38±0.01	**ND**	**0.31±0.07**	ND	ND	**ND**	**0.36±0.1**	**ND**	**0.38±0.03**
(*E,E*)-α-farnesene	1508	**0.05±0.04**	**0.69±0.1**	0.12±0.1	0.31±0.06	**0.03±0.02**	**1.22±0.03**	**0.68±0.1**	**4.58±0.7**	**0.18±0.1**	**0.66±0.06**

[1] Plants were treated 24 hours before the start of VOCs air entrainment. Under each chemical class, VOCs are ordered in accordance with their increasing retention time in a gas chromatograph and Kovats index. Bold values indicate significant differences between treatments (t-test; $P < 0.05$). # [DMNT: (E)-4,8-dimethyl-1,3,7-nonatriene; MeSA: Methyl salicylate; ND: Not Detected]. VOCs were tentatively identified based on spectra, Kovats retention index and NIST 17 library matches. KI: Kovats index determined on the intermediately non-polar HP5-MS column (https://webbook.nist.gov/; http://www.pherobase.com/).

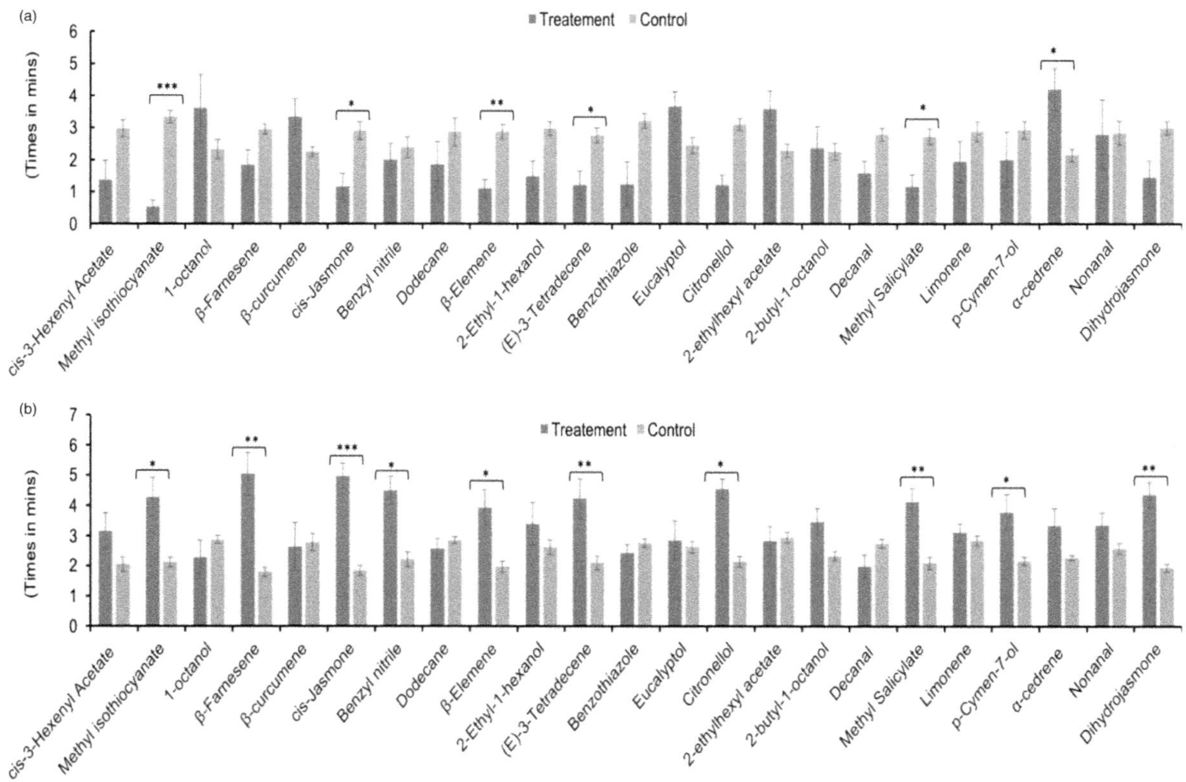

FIGURE 7.13 Behavioral responses of *M. persicae* (a) and *D. rapae* females (b) to standard compounds. Individual aphids were given 12 minutes to make a choice between one arm treated with synthetic compound vs. three solvent (Hexane) arms. The shown values are mean time spent in arm ± standard error (SE) ($n = 10$). Asterisks denote differing levels of statistical significance: * < 0.05, ** < 0.01 and *** < 0.001 (paired t-test) (Treatment; synthetic compound, control; Hexane).

α-cedrene ($p = 0.04$) was the only compound to which aphids were attracted and spent more time in the arm treated with synthetic compound. The other 17 compounds had no significant effect on aphids.

The results obtained from behavioral bioassay using synthetic analogues of identified compounds (Figure 7.13b) show time spent by parasitoids in the treated region of olfactometer compared to control when each compound was tested individually at the standard concentration of 100 ng. Of 23 identified volatile compounds, parasitoids spent significantly longer time in the arms treated with the following ten individual compounds; *cis*-Jasmone, citronellol, dihydrojasmone, methyl salicylate, (*E*)-3-tetradecene, β-farnesene, p-cymen-7-ol, β-elemene, methyl isothiocyanate, and benzyl nitrile. Of these ten compounds, *cis*-Jasmone ($p = 0.0004$), citronellol ($p = 0.0007$), dihydrojasmone ($p = 0.001$), were the compounds with highest attractant effect on parasitoids while methyl isothiocyanate ($p = 0.0258$), and benzyl nitrile ($p = 0.0102$) had the least significant effect on *D. rapae*.

7.5.3 Discussion

The total identified compounds in treated and untreated brassica plant volatiles were 24 in number. In this section, we tested synthetic analogues of all identified compound on an individual basis. The compound DMNT was not tested because of the unavailability of the compound. To test the effect of these compounds on insect behavior, behavioral bioassays were performed using a four-arm olfactometer and each compound was tested in a separate bioassay. Behavioral bioassay revealed that out of 23 compounds, six compounds had a significant effect on aphid behavior. Of these six, five compounds were repellent and one was attractive. Plant volatile benzenoids (methyl salicylate), ketones (*cis*-Jasmone), aliphatic hydrocarbons ((*E*)-3-tetradecene), and terpenes (β-elemene, α-cedrene) all induced significant changes in *M. persicae* behavior. On the other hand, parasitoid *D. rapae* showed significant attraction to ten out of the 23 compounds. *D. rapae* showed significant preference for terpenes (Citronellol, β-elemene, (*E,E*)-α-farnesene, p-cymen-7-ol), Ketones (*cis*-Jasmone, dihydrojasmone), nitrogen-containing compounds (methyl isothiocyanate, benzyl nitrile), benzenoids (methyl salicylate) and aliphatic compounds ((E)-3-tetradecene).

The compounds identified from CJ-treated plant volatiles, in the current study, have been tested against various species of herbivore pests in previous studies. These studies revealed that the induced volatiles compounds have potential to negatively affect plant pests, for instance; (*E*)-3-tetradecene acts as a repellent for *Tricholusia ni* and *Epilachna varivestis* (Liu, Norris and Marti, 1988; Liu, Norris and Lyne, 1989; Deepak *et al.*, 2019); methyl isothiocyanate causes mortality of *Otiorhynchus saculatus* (Borek *et al.*, 1997); dihydrojasmone derivatives affect the foraging activity of *M. persicae* (Paprocka *et al.*, 2018); limonene was effective against whiteflies, mealybugs, and scale insects (Hollingsworth, 2005; Conboy *et al.*, 2019); citronellol has been found to be responsible for oviposition reduction of herbivorous pests leafhopper, *Monochamous alternatus* and *Delia radicum* (Saxena and Basit, 1982; Klocke, Darlington and Balandrin, 1987; Lamy *et al.*, 2017); β-elemene as a constituent of essential oils (*Tetradium glabrifolium*, *Evodia rutaecarpa* and *Zanthoxylum rhoifolium*) showed strong repellency and larvicidal activity against Asian tiger mosquito (*Aedes albopticus*), *Triboilum castaneum, Lesioderma serricorne Liposcelis bostrychophila and Bemisia tabaci* (Christofoli *et al.*, 2015; Liu *et al.*, 2015; Cao *et al.*, 2018). (*E,E*)-α-Farnesene induces changes in behaviour of the codling moth (Sutherland, Hutchins and Wearing, 1974; Yan *et al.*, 2003) and p-Cymen-7-ol was found to be behaviorally active compounds when tested against tephritid pests and spring aphid *Kaburagia rhusicola* (Aluja *et al.*, 2020; Zhu *et al.*, 2020).

Furthermore, the compounds released with the CJ treatment of plants were also effective when tested with biocontrol agents i.e., predators and parasitoids. For example, CJ-treated tobacco plants release *cis*-3-hexenyl acetate was attracted to endoparasitoid *Campoletis chloridae* (Sun *et al.*, 2020), CJ treatment of soybean plants was effective in recruiting the stink bug egg parasitoid *Telenomus podisi* (Moraes *et al.*, 2009); an abundance of predatory thrips *Aeolothrips intermedius*, *Chrysoperla carnea* was observed in the cotton field treated with CJ (Bayram and Tonğa, 2018).

The overall number of compounds that had a significant effect on insect behavior was low compared to the overall number of identified compounds. While it is not unexpected that only a minority of compounds are bioactive, a further explanation for some compounds not showing a significant effect (repellent or attractant) could be the lack of appropriate mixture of compounds as these compounds were tested individually. Interestingly, compounds that had a significant effect on insect behavior were found in the brassica plants treated with CJ only. It has been reported that in addition to quality and quantity of VOCs, the combination and mixture of VOCs is also vital for inducing a change in insect behavior (Bruce and Pickett, 2011). Insects respond to plant volatiles with the help of a highly sensitive olfactory system that is made up of olfactory receptors neurons (ORNs) (Masson and Mustaparta, 1990; Bruce, Wadhams and Woodcock, 2005; Bruce and Pickett, 2011). Blends of plant volatiles play an important role for insects to recognise host and non-hosts and in making a behavioral decision that could be either avoidance or attraction (Wei *et al.*, 2007; Bruce and Pickett, 2011; Cunningham, 2012). The insect olfactory system can detect a single molecule compound, but it sometimes needs a whole volatile blend to recognise a plant as a host or non-host (Webster *et al.*, 2010). Individual compounds could be insufficient to induce a behavioral changes because of a lack of the combination of compounds present in the whole volatile blends (Beyaert and Hilker, 2014). That could be the reason for fewer compounds responsible for significant behavioral change in *M. persicae* and *D. rapae*, because the behavioral activity induced by blends of volatiles were higher when insects were exposed to whole blends. However, the key aspect here was to characterize the bioactivity of compounds induced by *cis*-Jasmone. Compounds that repelled aphids and attracted parasitoids were successfully identified. The identification and characterization of the *cis*-Jasmone-induced compounds that repel pests and attract their natural enemies can be used in developing new pest management strategies.

7.6 DISCUSSION

Data obtained from the GC-MS analysis revealed changes in the emitted compounds in blends of treated and control plants. CJ-treated plants released a richer blend of volatiles compared to control plants. There was a significant increase in the number of emitted compounds as well as the amount produced. CJ-treated plants released a volatile blend rich in volatile compounds such as (*E*)-3-tetradecene, (*Z*)-3-hexenyl acetate, 2-ethyl-1-hexanol, methyl isothiocyanate, dihydrojasmone, limonene, citronellol, methyl salicylate (MeSA), β-elemene, (*E,E*)-α-Farnesene and p-Cymen-7-ol. In addition, CJ production is also induced by CJ treatment and is coming from the plant. This is not slow release from the treatment itself because there would not be so much difference between the different cultivars if it was.

The effect of compounds identified from CJ-treated plant volatiles has already been tested against herbivorous pests. Previous studies revealed that these induced volatile compounds have the potential to negatively impact plant pests, for instance; (*E*)-3-tetradecene acts as a repellent for *Tricholusia ni* and *Epilachna varivestis* (Liu, Norris and Marti, 1988; Liu, Norris and Lyne, 1989; Deepak *et al.*, 2019); methyl isothiocyanate causes mortality of *Otiorhynchus saculatus* (Borek *et al.*, 1997); dihydrojasmone derivatives affect the foraging activity of *M. persicae* (Paprocka *et al.*, 2018); limonene was effective against whiteflies, mealybugs and scale insects (Hollingsworth, 2005; Conboy *et al.*, 2019); citronellol has been found responsible for oviposition reduction of herbivorous pests leafhopper, *Monochamous alternatus* and *Delia radicum* (Saxena and Basit, 1982; Klocke, Darlington and Balandrin, 1987; Lamy *et al.*, 2017); β-elemene as a constituent of essential oil (*Tetradium glabrifolium, Evodia rutaecarpa* and *Zanthoxylum rhoifolium*) showed strong repellency and larvicidal activity against Asian tiger mosquito (*Aedes albopticus*), *Triboilum castaneum, Lesioderma serricorne Liposcelis bostrychophila and Bemisia tabaci* (Christofoli *et al.*, 2015; Liu *et al.*, 2015; Cao *et al.*, 2018). (*E,E*)-α-Farnesene induces changes in the behavior of codling moth (Sutherland, Hutchins and Wearing, 1974; Yan *et al.*, 2003) and p-Cymen-7-ol was found to be behaviorally active compounds when tested against tephritid pests and spring aphid *Kaburagia rhusicola* (Aluja *et al.*, 2020; Zhu *et al.*, 2020).

Furthermore, compounds released with the CJ treatments of brassica lines were also effective when tested with biocontrol agents i.e., predators and parasitoids. For example, *cis*-3-hexenyl acetate attracted endoparasitoid *Campoletis chloridae* (Sun *et al.*, 2020). Treatment of field plots with MeSA attracted predator insects from families Braconidae, Empididae, Sarcophidae and ladybirds e.g. *Coccinella septempunctata* as well as parasitoids (*Chrysopa nigricornis*) of soybean aphid (James, 2003; James and Price, 2004; Mallinger, Hogg and Gratton, 2011). In behavioral bioassay females *Aphis craccivora, Lobesia botrana* showed preference to p-Cymen-7-ol (Katerinopoulos *et al.*, 2005; Mitra *et al.*, 2021). Some of these compounds have been found attractive to several plant pests too that can be useful in developing new management practices based on trap cards that will help in reducing pest burden on crop plants by mass trapping (Weinzierl *et al.*, 2005; Tewari *et al.*, 2014).

The chemical analysis of plant volatiles revealed significant increases in volatile emission from CJ-treated plants and provided evidence of both qualitative and quantitative changes. However, these changes in volatile emissions varied with genotype. An increased number of plant volatiles emitted was observed in genotypes Wesway and Pak choi when treated with CJ: the number of identified compounds was eight and ten in the blend released by the plants treated with blank formulation, but increased to 18 for both these genotypes after CJ treatment (genotypes Wesway and Pak choi respectively). Turnip rutabaga also showed an increase in the number of emitted compounds from 15 to 19. All five brassica genotypes had enhanced qualitative and quantitative emission of volatiles after exposure to CJ. Interestingly, compounds released by plants treated with CJ were responsible for inducing behavioral changes in insects (*M. persicae* and *D. rapae*) in the four-arm olfactometer. Based on the evidence from chemical analyses, and insect bioassay data, it is clear that CJ treatment induces plant defense against insect pests.

REFERENCES

Ali, J. *et al.* (2021) 'Effects of cis-jasmone treatment of brassicas on interactions with *Myzus persicae* aphids and their parasitoid *Diaeretiella rapae*', *Frontiers in Plant Science*, 12. doi: 10.3389/fpls.2021.711896

Ali, J. *et al.* (2023) 'Exogenous application of methyl salicylate induces defence in brassica against peach potato aphid *Myzus persicae*', *Plants*, 12(9), p. 1770.

Aluja, M. *et al.* (2020) 'Host plant and antibiotic effects on scent bouquet composition of Anastrepha ludens and Anastrepha obliqua calling males, two polyphagous tephritid pests', *Insects*, 11(5), p. 309.

Aranega-Bou, P. *et al.* (2014) 'Priming of plant resistance by natural compounds. Hexanoic acid as a model', *Frontiers in Plant Science*, 5, p. 488.

Arimura, G., Kost, C. and Boland, W. (2005) 'Herbivore-induced, indirect plant defences', *Biochimica et Biophysica Acta (BBA)-Molecular and Cell Biology of Lipids*, 1734(2), pp. 91–111.

Ballhorn, D. J., Kautz, S. and Schädler, M. (2013) 'Induced plant defense via volatile production is dependent on rhizobial symbiosis', *Oecologia*, 172(3), pp. 833–846. doi: 10.1007/s00442-012-2539-x

Balmer, A. *et al.* (2015) 'The "prime-ome": towards a holistic approach to priming', *Trends in Plant Science*, 20(7), pp. 443–452.

Bayram, A. and Tonğa, A. (2018) 'cis-Jasmone treatments affect pests and beneficial insects of wheat (Triticum aestivum L.): the influence of doses and plant growth stages', *Crop Protection*, 105, pp. 70–79.

Bektas, Y. and Eulgem, T. (2015) 'Synthetic plant defense elicitors', *Frontiers in Plant Science*, 5(Jan.), p. 804. doi: 10.3389/fpls.2014.00804

Beyaert, I. and Hilker, M. (2014) 'Plant odour plumes as mediators of plant–insect interactions', *Biological Reviews*, 89(1), pp. 68–81.

Bi, H. H. *et al.* (2007) 'Rice allelopathy induced by methyl jasmonate and methyl salicylate', *Journal of Chemical Ecology*, 33(5), pp. 1089–1103.

Birkett, M. A. *et al.* (2000) 'New roles for cis-jasmone as an insect semiochemical and in plant defense', *Proceedings of the National Academy of Sciences of the United States of America*, 97(16), pp. 9329–9334. doi: 10.1073/pnas.160241697

Blassioli Moraes, M. C. *et al.* (2008) 'cis-Jasmone induces accumulation of defence compounds in wheat, Triticum aestivum', *Phytochemistry*, 69(1), pp. 9–17. doi: 10.1016/j.phytochem.2007.06.020

Borek, V. *et al.* (1997) 'Toxicity of rapeseed meal and methyl isothiocyanate to larvae of the black vine weevil (Coleoptera: Curculionidae)', *Journal of Economic Entomology*, 90(1), pp. 109–112.

Bruce, T. J.A. and Pickett, J. A. (2011) 'Perception of plant volatile blends by herbivorous insects - finding the right mix', *Phytochemistry*, 72(13), pp. 1605–1611. doi: 10.1016/j.phytochem.2011.04.011

Bruce, T. J. A., Wadhams, L. J. and Woodcock, C. M. (2005) 'Insect host location: a volatile situation', *Trends in Plant Science*, 10(6), pp. 269–274.

Bruce, T. J. A. *et al.* (2003) 'cis-Jasmone treatment induces resistance in wheat plants against the grain aphid, sitobion avenae (Fabricius) (Homoptera: Aphididae)', *Pest Management Science*, 59(9), pp. 1031–1036. doi: 10.1002/ps.730

Bruce, T. J. A. *et al.* (2008) '*cis*-Jasmone induces Arabidopsis genes that affect the chemical ecology of multitrophic interactions with aphids and their parasitoids', *Proceedings of the National Academy of Sciences of the United States of America*, 105(12), pp. 4553–4558. doi: 10.1073/pnas.0710305105

Bruce, T. J. A. *et al.* (2017) 'Prospects for plant defence activators and biocontrol in IPM – concepts and lessons learnt so far', *Crop Protection*, 97, pp. 128–134. doi: 10.1016/j.cropro.2016.10.003

Budenberg, W. J., Powell, W. and Clark, S. J. (1992) 'The influence of aphids and honeydew on the leaving rate of searching aphid parasitoids from wheat plants', *Entomologia Experimentalis et Applicata*, 63(3), pp. 259–264.

Cao, J.-Q. *et al.* (2018) 'Contact toxicity and repellency of the essential oils of Evodia lenticellata Huang and Evodia rutaecarpa (Juss.) Benth. leaves against three stored product insects', *Journal of Oleo Science*, 67(8), pp. 1027–1034.

Christofoli, M. *et al.* (2015) 'Insecticidal effect of nanoencapsulated essential oils from Zanthoxylum rhoifolium (Rutaceae) in Bemisia tabaci populations', *Industrial Crops and Products*, 70, pp. 301–308.

Cohen, Y. R. (2002) 'β-aminobutyric acid-induced resistance against plant pathogens', *Plant Disease*, 86(5), pp. 448–457.

Conboy, N. J. A. *et al.* (2019) 'Companion planting with French marigolds protects tomato plants from glasshouse whiteflies through the emission of airborne limonene', *PloS One*, 14(3), p. e0213071.

Cortesero, A. M., Stapel, J. O. and Lewis, W. J. (2000) 'Understanding and manipulating plant attributes to enhance biological control', *Biological Control*, 17(1), pp. 35–49.

Cunningham, J. P. (2012) 'Can mechanism help explain insect host choice?', *Journal of Evolutionary Biology*, 25(2), pp. 244–251.

da Graça, J. P. *et al.* (2016) 'The natural plant stress elicitor cis-jasmone causes cultivar-dependent reduction in growth of the stink bug, Euschistus heros and associated changes in flavonoid concentrations in soybean, Glycine max', *Phytochemistry*, 131, pp. 84–91.

Dabrowska, P. and Boland, W. (2007) 'iso-OPDA: an early precursor of cis-Jasmone in plants?', *ChemBioChem*, 8(18), pp. 2281–2285. doi: 10.1002/cbic.200700464

D'Alessandro, M. and Turlings, T. C. J. (2006) 'Advances and challenges in the identification of volatiles that mediate interactions among plants and arthropods', *Analyst*, 131(1), pp. 24–32.

Deepak, P. *et al.* (2019) 'Phytochemical profiling of marine red alga, Halymenia palmata and its bio-control effects against Dengue Vector, Aedes aegypti', *South African Journal of Botany*, 121, pp. 257–266.

Delaney, K. J. *et al.* (2013) 'Synthetic cis-jasmone exposure induces wheat and barley volatiles that repel the pest cereal leaf beetle, *Oulema melanopus* L', *Journal of Chemical Ecology*, 39(5), pp. 620–629.

Dewhirst, S. Y. *et al.* (2012) 'Activation of defence in sweet pepper, Capsicum annum, by cis-jasmone, and its impact on aphid and aphid parasitoid behaviour', *Pest Management Science*, 68(10), pp. 1419–1429. doi: 10.1002/ps.3326

Dicke, M. and Baldwin, I. T. (2010) 'The evolutionary context for herbivore-induced plant volatiles: beyond the "cry for help"', *Trends in Plant Science*, 15(3), pp. 167–175.

Disi, J. O. *et al.* (2017) 'cis-Jasmone primes defense pathways in tomato via emission of volatile organic compounds and regulation of genes with consequences for Spodoptera exigua oviposition', *Arthropod-Plant Interactions*, 11(4), pp. 591–602.

Dudareva, N. and Pichersky, E. (2008) 'Metabolic engineering of plant volatiles', *Current Opinion in Biotechnology*, 19(2), pp. 181–189.

Effah, E., Holopainen, J. K. and McCormick, A. C. (2019) 'Potential roles of volatile organic compounds in plant competition', *Perspectives in Plant Ecology, Evolution and Systematics*, 38, pp. 58–63.

Egger, B. and Koschier, E. H. (2014) 'Behavioural responses of Frankliniella occidentalis Pergande larvae to methyl jasmonate and cis-jasmone', *Journal of Pest Science*, 87(1), pp. 53–59.

El-Sayed, A. M. *et al.* (2009) 'Attraction of New Zealand flower thrips, Thrips obscuratus, to cis-jasmone, a volatile identified from Japanese honeysuckle flowers', *Journal of Chemical Ecology*, 35(6), pp. 656–663. doi: 10.1007/s10886-009-9619-3

Esnault, D. *et al.* (2005) 'Natural plant defence mechanisms, laminarin and glume blotch resistance', *Phytoma*, (587), pp. 49–51.

Farmer, E. E. and Ryan, C. A. (1990) 'Interplant communication: airborne methyl jasmonate induces synthesis of proteinase inhibitors in plant leaves', *Proceedings of the National Academy of Sciences of the United States of America*, 87(19), pp. 7713–7716.

Giron, D. *et al.* (2016) 'Insect-induced effects on plants and possible effectors used by galling and leaf-mining insects to manipulate their host-plant', *Journal of Insect Physiology*, 84, pp. 70–89.

Giunti, G. *et al.* (2018) 'VOC emissions influence intra-and interspecific interactions among stored-product Coleoptera in paddy rice', *Scientific Reports*, 8(1), pp. 1–9.

Gols, R., Posthumus, M. A. and Dicke, M. (1999) 'Jasmonic acid induces the production of gerbera volatiles that attract the biological control agent Phytoseiulus persimilis', *Entomologia Experimentalis et Applicata*, 93(1), pp. 77–86.

Gouinguené, S. P. and Turlings, T. C. J. (2002) 'The effects of abiotic factors on induced volatile emissions in corn plants', *Plant Physiology*, 129(3), pp. 1296–1307.

Guenzi, W. D. and McCalla, T. M. (1966) 'Phenolic acids in oats, wheat, sorghum, and corn residues and their phytotoxicity 1', *Agronomy Journal*, 58(3), pp. 303–304.

Hare, J. D. (2011) 'Ecological role of volatiles produced by plants in response to damage by herbivorous insects', *Annual Review of Entomology*, 56, pp. 161–180. doi: 10.1146/annurev-ento-120709-144753

Hegde, M. *et al.* (2012) 'Aphid antixenosis in cotton is activated by the natural plant defence elicitor cis-jasmone', *Phytochemistry*, 78, pp. 81–88.

Heil, M. (2004) 'Induction of two indirect defences benefits Lima bean (Phaseolus lunatus, Fabaceae) in nature', *Journal of Ecology*, 92(3), pp. 527–536. doi: 10.1111/j.0022-0477.2004.00890.x

Heil, M. (2014) 'Herbivore-induced plant volatiles: targets, perception and unanswered questions', *New Phytologist*, 204(2), pp. 297–306.

Hokkanen, H. M. T. and Lynch, J. M. (2003) *Biological control: benefits and risks*. Cambridge University Press.

Hollingsworth, R. G. (2005) 'Limonene, a citrus extract, for control of mealybugs and scale insects', *Journal of Economic Entomology*, 98(3), pp. 772–779.

Holopainen, J. K. *et al.* (2009) 'Potential for the use of exogenous chemical elicitors in disease and insect pest management of conifer seedling roduction', *The Open Forest Science Journal*, 2(1), pp. 17–24.

Howarth, F. G. (1991) 'Environmental impacts of classical biological control', *Annual Review of Entomology*, 36(1), pp. 485–509.

Ighodaro, O. M. and Akinloye, O. A. (2018) 'First line defence antioxidants-superoxide dismutase (SOD), catalase (CAT) and glutathione peroxidase (GPX): Their fundamental role in the entire antioxidant defence grid', *Alexandria Journal of Medicine*, 54(4), pp. 287–293.

Jakab, G. *et al.* (2001) 'β-Aminobutyric acid-induced resistance in plants', *European Journal of Plant Pathology*, 107(1), pp. 29–37.

James, D. G. (2003) 'Field evaluation of herbivore-induced plant volatiles as attractants for beneficial insects: Methyl salicylate and the green lacewing, chrysopa nigricornis', *Journal of Chemical Ecology*, 29(7), pp. 1601–1609. doi: 10.1023/A:1024270713493
James, D. G. and Price, T. S. (2004) 'Field-testing of methyl salicylate for recruitment and retention of beneficial insects in grapes and hops', *Journal of Chemical Ecology*, 30(8), pp. 1613–1628.
Katerinopoulos, H. E. *et al.* (2005) 'Composition and insect attracting activity of the essential oil of Rosmarinus officinalis', *Journal of Chemical Ecology*, 31(1), pp. 111–122.
Kessler, A. and Baldwin, I. T. (2001) 'Defensive function of herbivore-induced plant volatile emissions in nature', *Science*, 291(5511), pp. 2141–2144. doi: 10.1126/science.291.5511.2141
Khan, Z. R. *et al.* (2008) 'Chemical ecology and conservation biological control', *Biological Control*, 45(2), pp. 210–224.
Klocke, J. A., Darlington, M. V. and Balandrin, M. F. (1987) '1, 8-Cineole (Eucalyptol), a mosquito feeding and ovipositional repellent from volatile oil ofHemizonia fitchii (Asteraceae)', *Journal of Chemical Ecology*, 13(12), pp. 2131–2141.
Lamy, F. C. *et al.* (2017) 'Artificially applied plant volatile organic compounds modify the behavior of a pest with no adverse effect on its natural enemies in the field', *Journal of Pest Science*, 90(2), pp. 611–621.
Leather, S. R. (2017) '"Ecological Armageddon"-more evidence for the drastic decline in insect numbers', *Annals of Applied Biology*, 172(1), pp. 1–3.
Li, B. *et al.* (2013) 'Antibacterial activity of two chitosan solutions and their effect on rice bacterial leaf blight and leaf streak', *Pest Management Science*, 69(2), pp. 312–320. doi: 10.1002/ps.3399
Liu, S., Norris, D. M. and Marti, E. (1988) 'Behavioral responses of female adult Trichoplusia ni to volatiles from soybeans versus a preferred host, lima bean', *Entomologia Experimentalis et applicata*, 49(1–2), pp. 99–109.
Liu, S. H., Norris, D. M. and Lyne, P. (1989) 'Volatiles from the foliage of soybean, Glycine max, and lima bean, Phaseolus lunatus: their behavioral effects on the insects Trichoplusia ni and Epilachna varivestis', *Journal of Agricultural and Food Chemistry*, 37(2), pp. 496–501.
Liu, X. C. *et al.* (2015) 'Larvicidal activity of the essential oil from Tetradium glabrifolium fruits and its constituents against Aedes albopictus', *Pest Management Science*, 71(11), pp. 1582–1586.
López-Gresa, M. P. *et al.* (2019) 'Effect of Benzothiadiazole on the metabolome of tomato plants infected by citrus exocortis viroid', *Viruses*, 11(5), p. 437.
Lou, Y. and Baldwin, I. T. (2003) 'Manduca sexta recognition and resistance among allopolyploid Nicotiana host plants', *Proceedings of the National Academy of Sciences of the United States of America*, 100, pp. 14581–14586. doi: 10.1073/pnas.2135348100
Lou, Y.-G. *et al.* (2005) 'Exogenous application of jasmonic acid induces volatile emissions in rice and enhances parasitism of Nilaparvata lugens eggs by theParasitoid Anagrus nilaparvatae', *Journal of Chemical Ecology*, 31(9), pp. 1985–2002.
Loughrin, J. *et al.* (1995) 'Volatiles emitted by different cotton varieties damaged by feeding beet armyworm', *Journal of Chemical Ecology*, 21(8), pp. 1217–1227.
Mallinger, R. E., Hogg, D. B. and Gratton, C. (2011) 'Methyl salicylate attracts natural enemies and reduces populations of soybean aphids (Hemiptera: Aphididae) in soybean agroecosystems', *Journal of Economic Entomology*, 104(1), pp. 115–124.
Mandour, N. S. *et al.* (2013) 'Effects of prohydrojasmon-treated corn plants on attractiveness to parasitoids and the performance of their hosts', *Journal of Applied Entomology*, 137(1–2), pp. 104–112.
Masson, C. and Mustaparta, H. (1990) 'Chemical information processing in the olfactory system of insects', *Physiological Reviews*, 70(1), pp. 199–245.

Mello, M. O. and Silva-Filho, M. C. (2002) 'Plant-insect interactions: an evolutionary arms race between two distinct defense mechanisms', *Brazilian Journal of Plant Physiology*, 14(2), pp. 71–81.
Mithöfer, A. and Boland, W. (2012) 'Plant defense against herbivores: chemical aspects', *Annual Review of Plant Biology*, 63, pp. 431–450. doi: 10.1146/annurev-arplant-042110-103854
Mitra, P. *et al.* (2021) 'Identification of Lathyrus sativus plant volatiles causing behavioral preference of Aphis craccivora', *Pest Management Science*, 77(1), pp. 285–299.
Moraes, M. C. B. *et al.* (2008) 'cis-Jasmone induces accumulation of defence compounds in wheat, Triticum aestivum', *Phytochemistry*, 69(1), pp. 9–17.
Moraes, M. C. B. *et al.* (2009) 'Attraction of the stink bug egg parasitoid Telenomus podisi to defence signals from soybean activated by treatment with cis-jasmone', *Entomologia Experimentalis et Applicata*, 131(2), pp. 178–188. doi: 10.1111/j.1570-7458.2009.00836.x
Najar-Rodriguez, A. J. *et al.* (2015) 'Aphid-deprivation from Brassica plants results in increased isothiocyanate release and parasitoid attraction', *Chemoecology*, 25(6), pp. 303–311. doi: 10.1007/s00049-015-0199-0
Naranjo, S. E., Ellsworth, P. C. and Frisvold, G. B. (2015) 'Economic value of biological control in integrated pest management of managed plant systems', *Annual Review of Entomology*, 60, pp. 621–645.
Niemeyer, H. M. (1988) 'Hydroxamic acids (4-hydroxy-1, 4-benzoxazin-3-ones), defence chemicals in the Gramineae', *Phytochemistry*, 27(11), pp. 3349–3358.
Obara, N., Hasegawa, M. and Kodama, O. (2002) 'Induced volatiles in elicitor-treated and rice blast fungus-inoculated rice leaves', *Bioscience, Biotechnology, and Biochemistry*, 66(12), pp. 2549–2559.
Oluwafemi, S. *et al.* (2013) 'Priming of production in maize of volatile organic defence compounds by the natural plant activator cis-jasmone', *PloS One*, 8(6), p. e62299. doi: 10.1371/journal.pone.0062299
Orr, D. (2009) 'Biological control and integrated pest management', in *Integrated pest management: innovation-development process*, Editors: Rajinder Peshin, Ashok K. Dhawan, Springer, pp. 207–239.
Paprocka, M. *et al.* (2018) 'Novel hydroxy-and epoxy-cis-jasmone and dihydrojasmone derivatives affect the foraging activity of the peach potato aphid *Myzus persicae* (Sulzer) (Homoptera: Aphididae)', *Molecules*, 23(9), p. 2362.
Park, S.-W. *et al.* (2007) 'Methyl salicylate is a critical mobile signal for plant systemic acquired resistance', *Science*, 318(5847), pp. 113–116.
Pedrol, N., González, L. and Reigosa, M. J. (2006) 'Allelopathy and abiotic stress', in *Allelopathy*, Editors: Manuel J. Reigosa, Nuria Pedrol, Luís González, Springer, pp. 171–209.
Peñaflor, M. and Bento, J. M. S. (2013) 'Herbivore-induced plant volatiles to enhance biological control in agriculture', *Neotropical Entomology*, 42(4), pp. 331–343.
Pettersson, J. (1970) 'An aphid sex attractant', *Insect Systematics & Evolution*, 1(1), pp. 63–73.
Pickett, J. A. *et al.* (2007a) 'cis-Jasmone as allelopathic agent in inducing plant defence', *Allelopathy Journal*, 19(1), pp. 109–118.
Pickett, J. A. *et al.* (2007b) 'Plant volatiles yielding new ways to exploit plant defence', *Chemical Ecology*, pp. 161–173. doi: 10.1007/978-1-4020-5369-6_11
Pirk, C. W. W. (2021) 'Exploring the kairomone-based foraging behaviour of natural enemies to enhance biological control: a review', *Frontiers in Ecology and Evolution*, 9, p. 143.
Rodriguez-Saona, C. R. and Frost, C. J. (2010) 'New evidence for a multi-functional role of herbivore-induced plant volatiles in defense against herbivores', *Plant Signaling & Behavior*, 5(1), pp. 58–60.
Room, P. M. (1990) 'Ecology of a simple plant–herbivore system. Biological control of Salvinia', *Trends in Ecology & Evolution*, 5(3), pp. 74–79.
Röse, U. S. R. and Tumlinson, J. H. (2004) 'Volatiles released from cotton plants in response to Helicoverpa zea feeding damage on cotton flower buds', *Planta*, 218, pp. 824–832.

Röse, U. S. R. and Tumlinson, J. H. (2005) 'Systemic induction of volatile release in cotton: How specific is the signal to herbivory?', *Planta*, 222, pp. 327–335. doi: 10.1007/s00425-005-1528-2

Saxena, K. N. and Basit, A. (1982) 'Inhibition of oviposition by volatiles of certain plants and chemicals in the leafhopperAmrasca devastons (distant)', *Journal of Chemical Ecology*, 8(2), pp. 329–338.

Silva, D. B. *et al.* (2017) 'Qualitative and quantitative differences in herbivore-induced plant volatile blends from tomato plants infested by either Tuta absoluta or Bemisia tabaci', *Journal of Chemical Ecology*, 43(1), pp. 53–65.

Smith, J. J. *et al.* (2009) 'Small molecule activators of SIRT1 replicate signaling pathways triggered by calorie restriction in vivo', *BMC Systems Biology*, 3(1), pp. 1–14.

Sobhy, I. S., Bruce, T. J. A. and Turlings, T. C. J. (2018) 'Priming of cowpea volatile emissions with defense inducers enhances the plant's attractiveness to parasitoids when attacked by caterpillars', *Pest Management Science*, 74(4), pp. 966–977. doi: 10.1002/ps.4796

Sobhy, I. S., Erb, M. and Turlings, T. C. J. (2015) 'Plant strengtheners enhance parasitoid attraction to herbivore-damaged cotton via qualitative and quantitative changes in induced volatiles', *Pest Management Science*, 71(5), pp. 686–693. doi: 10.1002/ps.3821

Sobhy, I. S. *et al.* (2014) 'The prospect of applying chemical elicitors and plant strengtheners to enhance the biological control of crop pests', *Philosophical Transactions of the Royal Society B: Biological Sciences*, 369(1639), p. 20120283. doi: 10.1098/rstb.2012.0283

Sobhy, I. S. *et al.* (2017) '*cis*-jasmone elicits aphid-induced stress signalling in potatoes', *Journal of Chemical Ecology*, 43(1), pp. 39–52. doi: 10.1007/s10886-016-0805-9

Sobhy, I. S. *et al.* (2020) 'Sensing the danger signals: cis-jasmone reduces aphid performance on potato and modulates the magnitude of released volatiles', *Frontiers in Ecology and Evolution*, 7(January), pp. 1–14. doi: 10.3389/fevo.2019.00499

Spinelli, F. *et al.* (2011) 'Emission and function of volatile organic compounds in response to abiotic stress', in *Abiotic stress in plants–mechanisms and adaptations*, Editors: Arun Shanker, B. Venkateswarlu, pp. 367–394.

Sun, Y. *et al.* (2019) 'An odorant receptor mediates the attractiveness of cis-jasmone to Campoletis chlorideae, the endoparasitoid of Helicoverpa armigera', *Insect Molecular Biology*, 28(1), pp. 23–34.

Sun, Y.-L. *et al.* (2020) 'The cotton bollworm endoparasitoid Campoletis chlorideae is attracted by cis-jasmone or cis-3-hexenyl acetate but not by their mixtures', *Arthropod-Plant Interactions*, 14(2), pp. 169–179.

Sutherland, O. R. W., Hutchins, R. F. N. and Wearing, C. H. (1974) 'The role of the hydrocarbon α-farnesene in the behaviour of codling moth larvae and adults', in *Experimental analysis of insect behaviour*, Editor: L. Barton-Browne, Springer, pp. 249–263.

Takabayashi, J. and Shiojiri, K. (2019) 'Multifunctionality of herbivory-induced plant volatiles in chemical communication in tritrophic interactions', *Current Opinion in Insect Science*, 32, pp. 110–117.

Tanaka, K. *et al.* (2009) 'Highly selective tuning of a silkworm olfactory receptor to a key mulberry leaf volatile', *Current Biology*, 19(11), pp. 881–890. doi: 10.1016/j.cub.2009.04.035

Tewari, S. *et al.* (2014) 'Use of pheromones in insect pest management, with special attention to weevil pheromones', in *Integrated pest management*, Editor: Dharam P. Abrol, Elsevier, pp. 141–168.

Thaler, J. S., Humphrey, P. T. and Whiteman, N. K. (2012) 'Evolution of jasmonate and salicylate signal crosstalk', in *Trends in plant science*. Elsevier Ltd, pp. 260–270. doi: 10.1016/j.tplants.2012.02.010

Turlings, T. C. J. and Erb, M. (2018) 'Tritrophic interactions mediated by herbivore-induced plant volatiles: Mechanisms, ecological relevance, and application potential', *Annual Review of Entomology*, 63, pp. 433–452. doi: 10.1146/annurev-ento-020117-043507

Turlings, T. C. J. and Ton, J. (2006) 'Exploiting scents of distress: the prospect of manipulating herbivore-induced plant odours to enhance the control of agricultural pests', *Current Opinion in Plant Biology*, 9(4), pp. 421–427.

Uefune, M., Ozawa, R. and Takabayashi, J. (2014) 'Prohydrojasmon treatment of lima bean plants reduces the performance of two-spotted spider mites and induces volatiles', *Journal of Plant Interactions*, 9(1), pp. 69–73.

Vasyukova, N. I. *et al.* (2001) 'Modulation of plant resistance to diseases by water-soluble chitosan', *Applied Biochemistry and Microbiology*, 37(1), pp. 103–109.

Wang, S. Y. and Zheng, W. (2005) 'Preharvest application of methyl jasmonate increases fruit quality and antioxidant capacity in raspberries', *International Journal of Food Science and Technology*, 40(2), pp. 187–195. doi: 10.1111/j.1365-2621.2004.00930.x

War, A. R. *et al.* (2011) 'Herbivore induced plant volatiles:Their role in plant defense for pest management', *Plant Signaling and Behavior*, 6(12), pp. 1973–1978. doi: 10.4161/psb.6.12.18053

War, A. R. *et al.* (2012) 'Mechanisms of plant defense against insect herbivores', *Plant Signaling and Behavior*, 7(10). doi: 10.4161/psb.21663

Webster, B. *et al.* (2010) 'Volatiles functioning as host cues in a blend become nonhost cues when presented alone to the black bean aphid', *Animal Behaviour*, 79(2), pp. 451–457. doi: 10.1016/j.anbehav.2009.11.028

Wei, J. *et al.* (2007) 'Plants attract parasitic wasps to defend themselves against insect pests by releasing hexenol', *PLOS One*, 2(9), p. e852.

Weinzierl, R. *et al.* (2005) 'Insect attractants and traps', ENY277, Avalaible: http://ufdcimages.uflib.ufl.edu/IR/00/00/27/94/00001/IN08000.pdf [Accessed 23 Maret 2015].

Wendehenne, D. *et al.* (1998) 'Benzothiadiazole, an inducer of plant defenses, inhibits catalase and ascorbate peroxidase', *Phytochemistry*, 47(4), pp. 651–657.

Wielkopolan, B. and Obrępalska-Stęplowska, A. (2016) 'Three-way interaction among plants, bacteria, and coleopteran insects', *Planta*, 244(2), pp. 313–332.

Yan, F. *et al.* (2003) 'Roles of α-farnesene in the behaviors of codling moth females', *Zeitschrift für Naturforschung C*, 58(1–2), pp. 113–118.

Yoshida, K. *et al.* (2021) 'Effects of Prohydrojasmon on the Number of Infesting Herbivores and Biomass of Field-Grown Japanese Radish Plants', *Frontiers in Plant Science*, 12, p. 695701.

Zhu, J. *et al.* (2021) 'Preparation of PdNPs doped chitosan-based composite hydrogels as highly efficient catalysts for reduction of 4-nitrophenol', *Colloids and Surfaces A: Physicochemical and Engineering Aspects*, 611, p. 125889.

Zhu, X. *et al.* (2020) 'Chemical composition and attractant activity of volatiles from rhus potaninii to the spring aphid kaburagia rhusicola', *Molecules*, 25(15), p. 3412.

Glossary

Accessions: A distinct, uniquely identified sample of seeds or plants, that is maintained as part of a germplasm collection from a single species
Alate: Having wings
Allelochemical: Chemicals exerts a detrimental physiological effect on individuals of another species
Allomone: Chemical affect the behavior of individuals of another species
Anholocyclic: A life cycle which completely lacks male insects
Apterous: Having no wings
Arrestant: Stimulus that stop movement
Attractant: Stimulus that attracts
Biotype: Morphologically indistinguishable subspecies that are different in behavior or physiology
Chlorosis: Yellowing of leaf tissue
Chronic: Persisting for a long time or constantly recurring
Constitutive defense: Constitutive defenses are always present in the plant and do not depend on the attack of herbivores
Deterrent: Stimulus that deters a particular behavior
Detriophagous: Feeds on detritus (particles of organic matter from the decomposition of plant and animal remains) or organic debris
Endosymbiont microbes: Microbes living inside the body
Episodic: Occurring occasionally and at irregular intervals
ETL: Economic threshold level
FAO: Food and Agriculture Organization of the United Nations
Fundatrix: Winged or wingless parthenogenetic female aphid produced on primary host from an egg
Gynoparae: Winged or wingless parthenogenetic aphid that produces sexual forms
Herbaceous plant: Any vascular plant that does not have true woody tissues
Heteroecious: Host alternating
Holocyclic: Exhibit sexual mode of reproduction
Honeydew: A sugar-rich sticky liquid, secreted by aphids and some scale insects as they feed on plant sap
Induced defenses: These defenses requires plant sensing (signals), such as wounding from herbivore attack as opposed to wounding from mechanical damage
Invasion: Biological invasion is a prominent example of colonization of new areas of suitable habitat.
Kairomone: Chemical that acts interspecies where receivers from another species take advantage of it
Monoecious: Non-host alternating
Morph: An individual of one particular form, as a worker ant, in a species that occurs in two or more forms

Mutation: A change in the DNA sequence of an organism
Necrosis: Cellular death results in tissue turning dry and brown to black
Olfactometer: A device used to measuring the sensitivity of odor
Oviparae: Sexual female that lays fertilized eggs after mating with male
Parthenogenesis: A reproductive strategy that involves the development of a female gamete without fertilization
Pheromone: Chemical compound which acts intraspecies
Phytochemicals: Chemicals produced by plants
Polymorphism: Presence of different forms (wing or wingless) of a insects in life cycle
Recession: A significant decline in economic activity that is spread across the economy and that lasts more than a few months
Repulsion: Stimulus has the tendency to move away the object from it
Semiochemical: A stimulus that conveys a message from one organism to another to modify the behavior of its recipient
Threshold level: The concentration or amount of a substance below which it cannot be detected
Virginoparae: Aphid produced by parthenogenesis

Index

Pages in *italics* refer to figures and pages in **bold** refer to tables.

For Product Safety Concerns and Information please contact our EU
representative GPSR@taylorandfrancis.com
Taylor & Francis Verlag GmbH, Kaufingerstraße 24, 80331 München, Germany

www.ingramcontent.com/pod-product-compliance
Ingram Content Group UK Ltd.
Pitfield, Milton Keynes, MK11 3LW, UK
UKHW021515250926
475702UK00043B/661
9781032509297